The Last Mile of INNOVATION

A Novel about the Challenges of Delivering the Right Solutions to Those Who Need Them

WILLIAM D. REED

Developmental Editor: Sarah Welch

ISBN: 979-8-218-86018-9

 This is a work of fiction. Names, characters, businesses, organizations, places, events, and incidents are either the product of the author's imagination or are used fictitiously. Any resemblance to actual persons or events is purely coincidental. For further information, you can contact info@visionpark.media.

The Last Mile of INNOVATION

VISION PARK MEDIA

To those who dare to listen intently to understand real human problems and transforms intention into action to ensure the right solutions find their way to the lives they were meant to change.

CONTENTS

PREFACE

T*he Last Mile of Innovation: A Novel About the Challenges of Delivering the Right Solutions to Those Who Need Them* is the first in the *People-Centric Innovation* series. Inspired by the business fable format, this novel and the ones that follow will allow readers to experience the end-to-end journey of building something useful for the world up close.

This is a book for those who dream of creating solutions for real-world people. For those who see problems, challenges, and unmet needs out in the world and decide to do something about them. These people go into the proverbial lab to build a product or service, then bring it to the marketplace to let others buy it and use it to solve a problem they're having.

This journey from forming an idea to getting it in the hands of people who need it is not easy. Often, it can be lonely. The market is littered with fantastic products, both with and without technological elements. On the surface, the product can look fantastic or the service can be extraordinary. However, that is no guarantee of market success.

For those innovating and building something fantastic in the lab that meets a real human need, it can be very frustrating when the world doesn't blink an eye. Some may never hear about your solution, and others may pass it by because they don't understand its value. Or more particularly, its value to them at that particular time.

Startups face this all the time, struggling to break in through all the market noise, and even established companies face challenges around trying new things that may distract them from their existing successes.

So, what does it take to have an idea, create something new, and make it available for purchase? It has often been said, "If you build it, they will come." In many cases, that's not true.

Once you create your new product or service, how will you let your target prospects know about it? In a very crowded marketplace, it can be very hard for your message to get through to let people know you're there. How will your marketing efforts elevate it above the noise? When you get in front of a prospect, how effectively will you or your sales team convey the value proposition in a way that resonates with your prospects?

What is the purpose of products and services?

That might seem to be a simple question for many who are looking to have their problems dealt with, but innovators sometimes miss the mark. What are cool products good for? Nothing, unless they meet the needs and wants of buyers.

Innovating for innovation's sake isn't the way. We must remain human in all of this. Our humanity must be expressed in the solutions we make available to the people we serve.

And yet, there's another layer, as well: Products and services; and their impacts on users, are delivered through the structure of a business.

A viable business model is necessary to ensure a solution can be designed and built properly, presented in the marketplace to let buyers know it's there, sold effectively to prospects, and supported on an ongoing basis. The right business model also empowers creators to stay relevant through continuous innovation. After all, the greatest solution ever created has limited value if the business collapses a year later and buyers can no longer access ongoing help and improvements.

All this is to say, we must challenge companies when they lean more heavily into features than the value provided to people using their solutions to address their problems.

Mired in feature-itis, we can lose track of the most important fact. *This solution is all designed to help real people.*

In this novel, we will go on a journey with an innovator as he navigates all of the challenges that arise between identifying a problem and helping real people solve it. We will better understand the challenges of starting a business that builds the right thing, gets it into the hands of those who need it, and operates on a business model that offers just the right balance of flexibility and sustainability.

The Last Mile of Innovation: A Novel about the Challenges of Delivering the Right Solutions to Those Who Need Them is the story of an innovator on his journey to help someone who is trying to change the world.

CHAPTER 1

A NEW PROSPECT

JACOB WAS STANDING IN LINE TO ORDER BREAKFAST FROM the food truck he thought made the best egg sandwiches in town. He had stepped out of his office in Innovation Hills; the hippest, coolest coworking space in the city, to go down to the park and clear his head.

If you were looking for innovative thinkers and creators, Innovation Hills and the park behind it, were the place to hang out and be inspired. It was nestled on the riverfront downtown. With an outdoor concert hall, river front, and lush green landscape, this space was made to think through tough problems and make breakthroughs.

"Why should you have to stop what you are doing to input the new information?" he was thinking to himself. This was the problem mucking up his morning. "The technology is available to help."

As the person in front of him in the line moved up, a man's voice behind him caught his attention.

"This is so frustrating," the voice said. "There has to be a better way to address this challenge!"

"We're not leveraging the shared insights we learn. That was the third member who had the same symptoms after having that procedure," the woman next to him said.

"But we know how to treat the underlying cause to head it off earlier," the man responded. "There's no reason our members have to endure that discomfort for so long before being treated."

It was Nia and Michael. Jacob knew them from seeing them around at various industry events, but he'd never officially met them. Nia owned WellThrive Health, a company that provided holistic health and wellness services, and Michael was the medical director.

"I wish I was there," Michael continued. "I could have intervened and saved him from having to go through that pain. I've written up an extensive treatment plan based on those symptoms. I don't know why nobody follows it."

"Technically that was in our internal knowledge base," Nia replied. "However, the portal's so clunky that nobody uses it. Most of the notes are in a separate document in our file repository. For all practical purposes, your treatment plan was buried. This isn't the first time something like this has happened." They paused for a moment, clearly frustrated, before Nia continued emphatically. "Our members deserve better from us."

At this, Jacob couldn't stop himself. He turned toward them.

"Excuse me, I didn't mean to eavesdrop. I heard part of your conversation. Hi, my name is Jacob."

He looked each of them in the eye before continuing. "I may have a solution that could address your challenge."

Taken aback, Michael replied, "Excuse me, sir?"

"I own a company that designs and builds complex solutions," he clarified. "You don't need to live with this problem any longer. Do you have a few minutes to chat?"

Nia and Michael looked at each other and agreed. After everyone got their food, they sat down at a nearby bench.

"I heard the challenges you were describing, and I've heard similar stories before," Jacob said. "Nothing on the market seemed to adequately fix the problem, so I was inspired to build a solution that would address it. But first, tell me more about your business."

Nia took the lead here. "Our mission is to improve people's whole selves: mind, body, and soul. People come to our clinics when they're frustrated with traditional health care. We provide exams, consultations, and wellness plans as part of a holistic approach to health." She took a bite of her sandwich and looked at Michael to continue.

"During consultations, our wellness advisors are required to input a lot of information about the member's current health, their medical histories, nutritional habits, and physical activities." Michael picked up where Nia had left off. "There are clues in all that information that we could use to identify familiar patterns and expedite the recommended wellness plan. Specifically, if we could get pain relief earlier, that would be game changer for our members."

"So, what's stopping you?" asked Jacob.

"As individual providers, we usually spot root causes pretty intuitively, but it's a lot more effective if we can look at patterns and treatment plans from similar cases. Some of that is captured in our records, but a lot of it is just in the heads of individual practitioners. We need it to be accessible to all providers," Michael said, then took a sip of his drink.

Nia nodded and took over. "Traditional medicine is reactionary; it's focused on isolating and managing existing symptoms and ailments. But we aim to be preventative. We're focused on intervening as early as possible to treat the whole person. In some ways, that means we have it harder. Looking at the whole requires more systemic documentation to connect the dots. If we can connect the right dots to identify early indicators, we can dramatically improve members' wellness."

"It sounds like you two have a passion for your mission," Jacob said. "I can feel it."

Michael turned and looked at Nia. "She is an amazing soul. When I met her, I could feel she was one with the universe," he explained, smiling. "We met at a holistic conference three years ago. I was looking for something new. Nia was the change agent the world needs."

"To say I'm a spiritual person would be an understatement," Nia confirmed. "Yes, and Michael had such passion for helping his patients. When I met him, he had hit the wall with the status quo in our healthcare system.

"I was through with passing the problem on," Michael said. "My patients were not getting better. There was so much pressure from the insurance providers, my business partners, regulators, etc. Everyone else's agenda came before my patients."

Nia picked the story back up. "And I knew I needed our practice to be grounded in medicine. I didn't want to be just a cliché, touchy-feely wellness center that ignores the practicality of medicine. Michael brings that side of the equation. I hired him on to run all my health medical services, and he challenges me in a good way. Traditional meets modern. WellThrive Health started with one clinic and grew to multiple branches both here and across the region. And we don't call people patients; they are members of our wellness family."

Jacob was amazed. "That is such an amazing story, you two!" He said around a bite of sandwich. Swallowing, he switched gears. "So, where's the sticking point?"

"It was last week that our documentation problem really shined a light on the missed opportunities it was causing. One of our members came in complaining about soreness in his leg. He was one of our regulars, and one of our nutritionists had recently changed his meal plan. He was not due for a physical exam for another six months, but his primary advisor noticed something in his posture during his stretching routine that she thought was unusual. Her colleague had seen something similar. The member's mood was changing, too, and he reported problems getting a good night sleep recently," Nia explained. "But nobody was able to connect the dots, and the pain stuck around until the member ended up going back to his previous primary care doctor, who then sent him to an orthopedic specialist."

"But this is the frustrating thing," added Michael. "We've had cases exactly like this before, and we've treated them successfully. *I've* treated them successfully, and I've written up the proven treatment recommendations for this group of symptoms, but nobody else had access to it in our database."

"This isn't an isolated issue though," Nia added. "Every time I visit one of our clinics, I'm amazed by the amount of knowledge our talented group of professionals captures every single day about holistic health and the interconnectedness of mind and body. But we've never had a good way to share all that information among all of our clinics." She took a sip of water before continuing. "When Michael told me about that specific case, the lightbulb went off. His team could have used what Michael had already previously learned." She shook her head. "We could've helped our member a lot quicker, and without surgery, if we had fixed that gap."

Jacob nodded. He suspected his solution was the perfect fit for Nia and Michael, but he wanted to learn a little more first. "Have you tried anything to fix that gap in the past?"

Nia popped the last bite of her sandwich into her mouth and gestured at Michael to field Jacob's question. "We tried more technological solutions but ran into challenges integrating with our existing processes and tools. Some were more focused on traditional health care practices and their specific systems. Others were geared toward the major EHR platforms and integrations that just weren't compatible with our processes. We ended up having to come up with tons of workarounds while searching for other alternative solutions."

"And all while trying to expand your clinics across the region," Jacob cut in, nodding with sympathy for Nia and Michael, who he now officially considered prospective clients.

Nia nodded emphatically, and Michael went on. "Various vendors provide different pieces of the puzzle, but early intervention wasn't possible because no one had the full picture. Preventative healthcare can save patients from more critical and corrective care, which has worse health outcomes and costs more money to patients, insurers, and society. But we've learned the hard way how fragmented data systems can get in the way of making life-saving decisions."

"Better interoperability and data-sharing could have enabled a timely, coordinated intervention," Jacob said, echoing what he'd heard Michael say.

"We try to share notes as much as possible, but those manual ways don't scale, especially when are team is distributed geographically," Nia explained. "I want to be on the front end of reducing harm and improving outcomes," she continued proudly. "I have two missions: First, treat the whole person up front, and second, lead transformation in the healthcare industry to provide better care when the body gets sick."

"But Nia's vision is much larger than what our technology can support right now," Michael added. "We don't have a traditional EHR system. We have some custom-built tools we use until we can find the right solution on the market. They serve our medical records requirements, but they're not doing enough to meet our standards. So we're in need of something that works for us."

Now that Jacob had a solid understanding of Nia and Michael's needs, he was finally ready to tell them about his solution. "Hey guys," he started, "let me tell you why what you've told me matches up to what I've seen in the field. This is why I started my company, Lucera and built a solution called TaskCore. It was born out of experience that you two are having right now. I believe that what I've built is what you've been looking for."

"We'd love to hear it," Nia replied.

"Okay, great. Let me dive into how my solution addresses your symptoms and solves the problem at its source. We map the customer journey and experience end-to-end. We develop a knowledge schema that we build from an in-depth observation of the human interactions of their journey. This schema reflects a systemic view of all the environmental elements of the defined experience of all the people and things involved in those shared moments together. We interact with each element in isolation, then make notes of all touchpoints and reflections. Those notes are objects, recording in various ways, like manual data entry, voice entry, or video observations. All those becomes elements of the schema." He balled up his napkin and tossed it into his empty sandwich bag. "Shared knowledge is available for each local interaction, and TaskCore has an engine that looks for patterns that can help identify intervention points. We use both human and machine inputs to identify opportunities."

"This sounds interesting, Jacob," Michael asked. "Is this specific to healthcare?"

"How would this work in our industry?" Nia followed up, nodding.

"TaskCore, is a general-purpose knowledge and workflow optimization engine," Jacob replied. "One that supports the user's specific industry, workflow, and culture. I design for the desired experience. In this case, we capture data in a way that makes insights available to the frontlines whenever it's needed. TaskCore can operate as the core knowledge management system for you, as a supplement to your existing applications."

Jacob spent the next hour going into the inner workings of the solution, adding that it runs on any smartphone, making provider access frictionless.

He got the expected feedback from both Nia and Michael: They smiled and sighed in relief that maybe somebody was actually hearing them for the first time; that their frustrations were being seen for what they really were, not filtered through a lens of foreign assumptions.

Nia said, "Oh, I see how using TaskCore in that way would leverage our collective knowledge." She told him about how they had tried many solutions before, how they'd sat through countless sales pitches and product demos but none had come close to what Jacob was describing at this moment.

Michael agreed, expressing optimism that finally, in this very noisy and crowded technology marketplace, their problem may have met its solution.

As Jacob collected his trash and prepared to return inside, Nia asked him about his origin story.

"So, Jacob, I'm curious: Where does your passion and focus come from?"

"My parents," said Jacob quickly. "That was an easy question. My dad is an architect with a focus on designing public buildings that cater to the visitors and workers in the space. And my mom is an internal medicine doctor. They both showed me every single day our responsibilities to make the world a better place."

"Oh, your mom is a doctor." Nia chuckled. "Your secret weapon Jacob."

"Oh, we solution designers are nothing without context." Jacob grinned. "Every morning over breakfast, my dad would ask me how I planned to help the world today." He smiled as he recalled sitting with his parents at the breakfast table, a stack of homemade chocolate chip pancakes in front of him. This Saturday morning ritual was a highlight of his childhood. "Answering that question every morning, and talking about my ideas with my parents, I learned early to start identifying problems and thinking about how to solve them."

Jacob admired both of his parents. They were so successful, there wasn't anything they couldn't tackle when they put their minds to it. He knew from a young age that, when he grew up, he would make a difference in other people's lives, just like they did.

"When I was in college, I had the chance to get a lot of experience doing ethnographic research. I enjoyed studying people in their daily lives," Jacob recalled. "I was working on a research project exploring different team dynamics and performance under tight deadlines. I was fascinated with the patterns I observed in how people shared information and how they applied what they learned. There was so much loss of insights and so many missed signals on a daily basis."

Jacob recalled how he took what he saw and went to the drawing board to experiment, explore, and ideate on these particular challenges. Ever since he was a kid, he's loved to take problems apart. While others may have seen symptoms, he was always hungry to find the root cause.

"I liked the five whys approach and drilling down to find different ways I could resolve the problem closer to its roots as opposed to being distracted by attacking the symptoms."

"Me too, just in a different context," Nia joked. "However, when I was in school, I would go down to seven levels of why. Pushing deeper seemed to make a difference in figuring out a better approach forward."

"Seven...wow, Nia! I love that commitment. I believe we design for the results and experiences we want. And when we have someone in front of us that we are serving, we have to bring all our training, experiences, and knowledge to bear to help solve real problems from real people," Jacob said.

As Jacob recalled his journey from school to starting his company, he could feel a grin gleaming over his face all on its own. He didn't mind. In fact, he hoped Nia and Michael noticed.

To him, it seemed like a problem just met its solution.

After twenty minutes, Jacob bid his goodbye to return to work, feeling good about the conversation.

Six weeks later, WellThrive was Lucera's newest client.

CHAPTER 2

THE TEAM AT WORK

JACOB CAME INTO HIS OFFICE IN INNOVATION HILLS EARLY TO catch up on synthesizing some user research he'd collected during a recent creative workshop with a client. His meeting with his partners was coming up soon, and he wanted to clear his mind for that.

Jacob always took partner meetings seriously, even though his partners were his two best friends, Rhonda and Tim. They went back a long time.

Rhonda had always been into electronics, engineering, and mathematics. Building things with her hands was her passion. Something about putting together programmable robots had excited her when she was a kid. She'd thought if she could put a motor on an object, she could make magic. To this day, Rhonda was great at putting things together.

Tim was a whiz coder. He could look at source code and repurpose it in a totally different programming language. He had a knack for debugging errors and performance issues in modules.

They'd gone into business eighteen months ago, and TaskCore was their flagship product.

"So, guys, I'm going to go over this roadmap here," said Tim once they were huddled up in Jacob's office. "I made significant progress on making TaskCore work better for healthcare professionals."

Jacob nodded. "That's good to hear, Tim. Talking with Nia and Michael about their use cases provided some good insights on some of the focus areas we need to be mindful of."

Tim dove in. "The AI-Recommender on the wellness plan will mine the existing nuggets of insights that the advisors plug into the member record and correlate secondary and tertiary symptoms with validated member feedback on positive results."

"Yeah, and I'm focusing in right now on this Bluetooth connectivity and this new tablet that we're going to use to supplement the smartphone app," Rhonda chimed in. "Looks like this will work well for the shared medium that we'll use between their advisors and members."

"Yes, Jacob said, "based off the way I've mapped out the workflow from our conversations, the input will be great on the phone itself. But just because of the small real estate size, I think the tablet's larger screen will provide better value here that they can use to blow up certain images of the member's records and to view their wellness fitness plans."

Rhonda said, "I think that will work."

Tim agreed. "Yeah, we have great flexibility here with the way we've done the code base. The modifications I'm looking at will really hit home for what they're looking to do. The large language modules we're using on the back end will stay the same. The AI-assist function on the form fields will also stay the same." Tim paused. "I did have to take a look from an external standpoint at enhancing some of the privacy requirements to protect the health data. I'm still waiting to get a response on the implications of filtering that type of data out to our third party who's doing the transcribing and encoding. I've got the fields on personal identifiable information tokenized, so we'll be good there. But we'll need some additional resources to help out here, Jacob."

"You think we'll be okay here?" Jacob asked Tim. "I know we've been positioning TaskCore as industry agnostic, but it looks like we do need to be more mindful of how this workflow looks in a healthcare environment. Specifically this one, which has a heavy wellness flavor to it. That's good for us since the requirements aren't as stringent as if this was in a hospital setting."

Rhonda said, "Yes, I think that does make a difference. We've had trouble with healthcare and clinics specifically going direct. And there were a lot of requirements to try to partner with someone else in that industry. But them being a wellness standalone provides a lot more flexibility for us to do what we need to do without as many regulatory requirements. So, the connection between our app and the tablet that they'll use will not be an issue. I think that looks good."

"That takes care of Nia and Michael," Jacob said. "I do want to go out next week and meet with some more people in some other industry verticals too, potentially retail and some hospitality industries. I've been getting a lot of calls about that recently.

"Hey, slow down there," Tim chimed in. "I'm already trying to nail down this health and wellness industry. Don't overload me now," he joked.

"I'm mindful of that," Jacob replied. "I just want to make sure we have enough at-bats here. I think we kind of anticipated we would need to make some changes sector by sector. I will try to get more healthcare and wellness to round that out. But I don't want to lose traction with other industries as well."

They finished up their roadmap discussion and made the necessary changes to address Nia and Michael's industry, then talked about some additional features to improve workflow. Rhonda also wanted to take a look at a product roadmap for additional devices that may be used in the future.

With that, the three wrapped up their session and got back to work.

As Jacob was driving home, his phone rang, and he saw it was coming from Danielle. She worked for a service company that specialized in building APIs, and she was his go-to when he needed integration work done on his platform. And she was great at reading his mind.

He picked up immediately.

"Hi, Danielle. How are you?"

"Hey, Jacob. I'm doing great. I saw your interview yesterday on *BizTech Roundup*. It was a fantastic discussion."

"Oh, thanks. I appreciate that," Jacob replied.

"What are you up to?" she asked, with a familiar tone in her voice that meant she knew Jacob was struggling with his platform.

"I'm going home right now. There's something that's been bugging our system here, and I need to figure out a way to work around the compatibility and regulatory scope," he confessed.

"I think I might have heard you mention something on the show!" Yep, she already knew. "My team and I think we've got something for you.

"Oh, wow. Danielle, what are you doing, reading my mind?"

"Not exactly, but based on the way you were talking about your pain point in the interview, we've got some ideas on how to address it. I think we can put something together to help with that compatibility issue between those three systems you've been talking about."

"Hmm. Interesting," Jacob said.

"We already got something mocked up, so if you come by later in the week, I think we can show it off to you," she continued.

"Danielle, that would be fantastic," Jacob replied, thrilled but not surprised. "You always know when I need something."

"That's why they pay us integration consultants the big bucks!" Danielle laughed, but Jacob knew she deserved every penny. Danielle didn't have much love for product development, but she was a whiz at integration. She and her team could listen to a product developer talk about an integration problem and identify how to build the needed connection points at lightning speed. It really was her calling, and Jacob was always grateful for her support.

Sure enough, when he met with Danielle the next day to review her idea, it was golden.

"So, when you mentioned that the compatibility issue was with the device interface coming into your back-end process instance, I was thinking there's a different way of actually spinning that up. You can go ahead and use containers, and then we'll use a platform-as-a-service to bring these two elements over. And we can do it in-memory as opposed to shipping it out via API. It's almost like writing that file out at a much higher speed. That way, it can go into a different ecosystem where your partners can more easily integrate inside the feature set. That should eliminate you having to deal with the compatibility on your side while the third-party platforms handle it for you." Jacob knew instantly that this would be the perfect solution, but like always, Danielle was a step ahead of him. "I went ahead and spun up a proposal for you, Jacob. Let me know if that works," she continued.

"Danielle, no need to wait for this. It works," he said quickly. "You just amaze me with your ability to have these discussions without an engineer with you. Hard to believe you are just in sales." He grinned, knowing full well that there was no "just" about sales, or Danielle's skill in the job. Danielle laughed and joked she didn't take offense.

"I need this, and this phased approach will work for me." Jacob indicated to the proposal.

"Well, great, Jacob. Get the paperwork back to me, and we'll get started here in the next couple of weeks," she said.

"Sounds good. Danielle. Thanks for making it so easy to work with your organization," Jacob said before heading out.

The next morning, Jacob's alarm went off, followed just seconds later by his cell phone ringing. Groggy, he blinked against the morning light pouring through his window, grabbing his phone without checking the Caller ID. "Good morning, this is Jacob."

"Hey, Jacob. This is Angela from Allora. Just need a few minutes here. I'm really, really excited to talk with you. I know I got something that you guys can use with your product. I've been following your path in the technology market, and I think our solution would be a perfect fit for yours."

Before he could even process what was going on, she had gotten out four long sentences. He groaned internally. Allora Technologies sold many products and services in the wearable tech field, and for the last couple of weeks, Angela had been persistent in calling Jacob's cell phone number in an attempt to sell his company something. She'd gotten his private cell number from a third-party marketing service that finds contact numbers for potential prospects.

And this morning, Angela had finally gotten lucky when Jacob picked up without thinking.

Angela was insistent that Jacob give her a few minutes. She was going to pitch him hard on a new product that she just knew he had to have. She didn't bother to ask any questions. She said she had researched him and his platform online and was convinced one of their new products would be a great fit.

Angela dived into all the benefits that he could tell she had rehearsed before.

Jacob had to cut her off. "Hey, Angela, I really appreciate your call, but right now is not a good time. I was—"

But Angela just talked over him. "Hey, I just need a few more minutes, and I can get you a great discount if you buy before the end of the month. When do you think we can demo our product for you and your team?"

Jacob, getting frustrated, tried to push back. He didn't know exactly what her solution was, but this definitely was not the time to go into detail. Jacob contemplated just hanging up as Angela kept on going down the benefits of her product.

"Is this what our market has come to?" he thought. He couldn't stand the constant blind solicitation through cold calling, email blasts, and online ads. Was there no better way to connect an offering to its potential prospects?

Jacob understood Angela's perspective. She was just trying to sell her product, and he was doing the same thing... just in different formats. When his company grew larger, he dreamed of the ability to have a team of Angelas on his side; though perhaps more seasoned Angelas, to call on potential prospects on behalf of his solution.

Still, being on the other side of the phone, his empathy was starting to wane. "There just has to be a better way."

Finally, Jacob shifted his thoughts back to the phone call and interrupted her. "Sorry, Angela. I really have to get going to a meeting. Why don't you send me an email with a link to your information? I'll take a look when I get the chance."

CHAPTER 3

LISTEN AND LEARN

JACOB TURNED UP THE VOLUME ON THE SPEAKER. ROBERT WAS about to be interviewed on the *New Tech Market Landscape* podcast. Jacob definitely wanted to check this out. He loved hearing Robert speak about market conditions, investments, venture capitalists, technology innovation, and market forces in general. When he spoke, everybody listened.

Robert was Jacob's mentor from many years ago, a bonus father figure in some ways. He and his father always used to joke that Robert was like the brother that his father never had. Robert was like family to him and provided a lot of insights as he was going through school and starting his own company. So, he was not going to miss this opportunity to hear some more wisdom from his pseudo-Uncle.

The interviewer's first question was about the general landscape and shape of the technology market today. Robert started out talking about the fact that many companies; small, mid-size, and large enterprises, were being challenged to do more with less, and how to leverage technology to be the next big thing in their innovation jump.

"Customers are looking for differentiation, innovation, and value. Companies must be able to meet that need and meet the customers where they are. Many organizations struggle to build this innovation, and even when they do, sometimes they struggle to break through all the market noise and get to those prospects who could really use what they're positioning."

But, he added, "The market is wide open for those who are thinking differently and have an innovative use for technology that will help solve real human problems. And that's what organizations must do. Startup founders and heads of large organizations must be focused on value creation. Everything from the front end of the organization, from sales and marketing all the way to the back end of customer support, must be designed to highlight market relevance and grow their customer base."

Robert emphasized that, as he tutors young startups, he encourages them to focus on finding real problems out in the world and building something to address that. "Sometimes," he said, "people are chasing a shiny new version of what someone else has already created and not thinking outside the box to go find something new."

"Sometimes founders are impatient, cashing out too quickly, giving away too much equity and control in exchange for capital, or chasing the last coolest thing," he explained to the panel. "And sometimes startups chase new features instead of real solutions to buyers problems."

And hour later, Robert wrapped his interview by recommending that everybody in the technology field think about innovating based off the perspective of end user needs and not their own.

Jacob clicked off the interview, hoping to one day have just a sliver of insights Robert had.

Robert's advice was still rattling around in Jacob's head as he navigated his way through the crowd at the *Future of UX Conference* and scanned his badge at a breakout session for wireframing for new design of immersive applications. This was the most prominent tech conference of the year, and Jacob always made sure to attend. It was always a great chance for him to catch up with friends and associates that he'd met over the years and to get new insights into what the user experience community is talking about and looking forward to. Jacob was a UX designer at heart.

In this particular breakout session, he was looking forward to getting the inside scoop on what's new in the design phase of new products. The speaker was a well-known UX thought leader who always provided real insight into approaches for future designs. At one point during the presentation, Jacob looked around, observing that the audience was just as enthralled as he was. When the Q&A session started, Jacob was the first to raise his hand.

"Yes, you there, sir, in the front."

Jacob stood up. "I'm curious about how you are doing the wireframing on that new online platform. I see that you're using some of the new AI features to help spread out the examples of what the app could look like. I'm interested in how you make sure you get the right workflow in that wireframe, with the approach you're taking."

"Great question," said the speaker. "What's your name?"

"Jacob," he replied trying not to let his pride be too obvious in his voice.

"Very good question, Jacob. This is one of the biggest things we see here in the UX community. How do we test our ideals and get feedback from potential users at a faster pace? This particular online platform that I'm using with these new artificial intelligence features is something I'm experimenting with."

He explained some of the hot topics about artificial intelligence in the UX field and on some of the platforms a lot of UX designers use. "Here we have different ways of getting the inputs in on particular layouts of the actual app," he said. "But then also see what that looks like on the web front end as well. So, you can take the web and mobile front-end looks on both iOS and Android. Use this feature here—" he clicked something on the screen "—and this will map out a recommended step-by-step, screen-by-screen process. And then down here—" he underlined another feature with his laser pointer "—you can track this with the user journey. And then you map each one of those points back to the actual pain points and benefits that the user will experience going through this."

The speaker expounded on it for a few more minutes as he took additional questions. Jacob took lots of notes. He'd been hearing more and more about some newer tools available to try out online, and he was always keen to see what was new that would help out in his field.

After the presentation, he caught up with the speaker, who answered his follow-up questions and also introduced him to Kirk, one of the product managers of Wireloom, the online wireframe tool he'd just demonstrated. Jacob exchanged some information with the project manager and added him on LinkedIn so he could keep an eye out for more resources in the future.

"How are you doing, Mom? How was work?" Jacob asked his mom over dinner at his parents' house that night.

"Doing good," Jacob's mom said. "It's been a long day. I saw a lot of patients and had to stay late to catch up on dictating and charting. Sometimes that's the hardest part of the day. I spend so much time entering information that it makes me feel rushed dealing with patients."

"Yeah, I know, Mom," Jacob said. He'd heard her talk about this many times. After she took furious shorthand notes in the exam room, she had to capture it all in the online system later, then fill out the rest of the information at the end of the day to enter it into the EHR. "That's tough," he said.

"I know the promise of EHR systems was to be able to capture all this faster, but there are still so many regulations and requirements around capturing so much data. It does detract from the experience with the patient, which is the best part of my day," she said.

"So, Mom, do you think the EHR will ever improve to help with that?" Jacob asked.

"Ah." Jacob's mom smiled at him. "You've got your UX design hat on now."

Jacob laughed. "I always have that hat on, Mom. You know, this sounds like a job for TaskCore."

His mother chuckled. "Potentially. Our hospital has so many large integrator and technology solution providers in their system already. They mentioned some roadmap where in the future they hope to improve that, especially with all the talk of artificial intelligence. But for now, they rely on the doctors to just grunt it out and figure out how to get the information into the system.

"TaskCore really helps out in a lot of human interactions. But I never really zoomed in on the healthcare industry. Why don't you tell me more about your workflow and the type of information you enter? You've got my mind racing on this." He saw this as an opportunity to bond with his mom *and* gather some information that might help him help WellThrive.

They spent the remainder of the night walking through a day in the life of a doctor.

CHAPTER 4

THE BIG BOYS GET ALL THE ATTENTION

As Jacob sat back in his chair with his feet on the table, he picked up the remote control to turn up the volume on the TV for CNBC's *Lunch Power Hour* show featuring the movers and shakers in business discussing new innovations and what was going on in the market.

Jacob was especially tracking the trends in leveraging artificial intelligence to improve knowledge workers' productivity. He was curious to see the interview with Sam Chambers, CEO of JunoTech. He'd known him in college. He'd always thought Sam promised more than he could deliver, but he was a great presenter and could sell almost anything.

And as expected, he stole the spotlight on the panel right away.

Jacob admired the show's panel for their business acumen and ability to identify the next big thing, but today they were fawning all over Sam.

"Doesn't exactly do what he says it does, but I'm game to listen", Jacob muttered to himself. He marveled at how much admiration the panel gave to Sam for something he knew had serious technical limitations behind the scenes.

"Look like you're preparing your company to be acquired," said one of the panelists. "You about to go big."

Sam chuckled. "Hey, you just got to follow me closely to see what's going to happen next."

Jacob had a sense that Sam would sell and then leverage someone else's team to try to make his product live up to the hype. In one sense, that was a shrewd business move, but on the other hand it bothered him that Sam was misleading people about what his solution could actually do.

Jacob thought it should be him on that show. His product actually worked as advertised; and he was going to prove it at TechSolve next week.

As he hit the floor at one of the tech industry's biggest conferences in the area, Jacob came full of energy. He couldn't wait to show the real value of his product to attendees. The conference attracted businesses and IT leaders in all types of industries to learn about the new technological innovations and solutions coming to market this year.

Jacob walked around the main hall the first afternoon to get the pulse of the attendees. He loved overhearing conversations from prospects to understand their views on their problems and the types of solutions they were seeing on the floor. He made sure he went to all the booths, interacting with attendees as they milled around the large exhibition floor and chatting with other sales and technical teams.

The startups and smaller players like Lucera were on the outside edge of the exhibition floor. A small folding table and an easel with their logo fought valiantly for people's attention as they walked around the floor. They didn't have any fancy swag or giveaways, or any flashy neon lights or displays. All he had was a story to tell; one Jacob knew the right prospects would benefit from hearing. But on this day, as on so many days, the big boys garnered most of the attendees' attention.

Standing at the table with Rhonda and Tim, plus several of their developers and their only sales rep, Jacob watched large groups bypassing their small table on their way to the bigger booths. These bigger booths were more established companies, much larger in size and scale, with bright, gleaming lights, lots of handouts, giveaways, and other seemingly irresistible prospect magnets.

The five biggest players had center stage in the middle of the exhibition floor.

"They get all the attention because they have big marketing budgets," Rhonda grumbled. "Still, though, I marvel at how many products they can offer their customers."

When someone stopped by Lucera's table, it seemed like they were more interested in getting their badge stamped to be included in a raffle for some prize. When they could garner real interest, it was always from folks who couldn't make, or even influence, buying decisions; or else they shied away from Lucera's small size and relatively unknown nature.

Jacob recalled the last person who stopped. "Yeah, man," the prospect had said. "I really appreciate this. It sounds like a great solution, but my company has a lot of paperwork you have got to go through to even get a new vendor in our system here. I can take your card and maybe get you in front of our team, but just to be truthful, you got a lot of bigger companies in front of you guys."

Another woman acknowledged that her company had the problem

Lucera was designed to fix, but because they had so many other solutions in place that weren't fully implemented, she didn't think she could get the budget for another tool at this time.

Then there was the visitor who who'd seemed more interested in the speeds and feeds and the technical wizardry of it all but not the potential business applications.

"What a waste. They are missing out," Tim said.

Jacob was disappointed that the Lucera table seemed like such an afterthought, but he tried to encourage his team to stay upbeat. They all knew they had something better than their peers. They just needed more chances to tell their story.

A few hours later, as Jacob roamed the hall again, someone from one of the large booths gestured over to him. It was a middle-aged man in his early forties in a nice tan suit. As Jacob walked toward the booth, he could see the company logo: Omnify Technologies.

"I'm Tom," said the man, shaking Jacob's hand enthusiastically. "Do you have a few minutes?"

"I do," Jacob responded. "You're getting so much foot traffic over here. From your displays looks like you guys play in a lot of different areas."

Tom nodded. "Sometimes you don't see companies like us at these conferences, but we play in healthcare, retail, education, and finance, to name a few."

"Yeah, that seems to give you an advantage. One of the challenges us smaller players have is just breaking in on some of these conversations."

"I had a chance to pick up one of your brochures from your table earlier. It appears TaskCore can be applied to many verticals as well." Tom gestured as he flipped through the pages. "You're not just pigeonholed to one industry. Like what we do, but with you guys being so small, you may have to focus on just one or two areas. Still, though, I'm curious about TaskCore. It looks like it really can help a lot of different sectors improve workplace collaboration. Tell me more." Tom inquired.

Jacob spent the next ten minutes explaining the ins and outs of TaskCore to Tom.

"There are people at this conference who should taking a look at your what company provides. Very intriguing solution." Tom said. He seemed genuinely interested in the solution.

Jacob relayed the challenges of even getting a meeting inside the health field vertical. He'd had success with other industries, like retail and tech and some services, but healthcare had too many upfront requirements to even get in the door.

"You're right, Jacob," Tom agreed. "There's a lot of red tape and regulation in healthcare. In a lot of cases, a lot of people have to work together to introduce even the idea of new technologies into hospital healthcare systems. And it's much more challenging for new upstarts like you."

Tom thanked him for his time and said he needed to get back to his booth duties but looked forward to hearing more great things from Lucera.

"And that's the problem with the industry. No matter what we do, we can't find somebody who can actually fix our problem. We run into the same issue year after year after year. We keep trying different products and solutions, and they all keep coming up short."

As Jacob milled around one of the conference's many networking events the next afternoon, he overheard yet another conversation. This time, an exasperated man in his mid-fifties was venting with another younger man about the problems they were experiencing as their company continued to fail to fix a persistent and painful problem.

"We have reorg after reorg, brainstorming session after brainstorming session," the older man continued. "We go out and do all these demos of all these different solutions that are claiming to help, but we get products that don't pass muster. Half of them can't make it through our evaluation trials before failing to miss the mark."

Here, the younger man chimed in, clearly eager to express his own frustration as well. "They've got a lot of interesting features, but nothing that actually moves the needle for us. We don't need cool features; we just want to get this damn problem fixed."

Jacob's ears perked up. Like he'd done with Nia and Michael, he interjected impulsively. "Excuse me, I couldn't help but overhear your conversation and the obvious frustration you guys are having. Can you tell me a little bit more about what you're experiencing and what you've tried?"

"Yeah," said the older guy. "Like I tell my buddy here, we keep throwing bodies at this problem. I just can't understand why we can't get some of the vendors out there to innovate just a little more to help with this. We just can't scale it. We can't pin down the problem, and our customers are feeling the pain. Vendors can't see where we are. We're struggling to come up with something that will actually work for them. But at the end of the day, our solution isn't working because we can't get this one feature to work right. We just cannot find anything on the market to help us out."

"Guys, I think there definitely is something out there," said Jacob. "My company's been in business for one reason and one reason only, and that's to fix this type of issue you were discussing."

The two guys looked skeptical. "Who are you again? Which company do you work for?"

"Oh, I'm the founder and CEO of Lucera," Jacob responded proudly. "TaskCore is our flagship solution."

"TaskCore. I've never heard of it," the older gentleman said, but the younger one pulled up Lucera's website on his phone and showed the screen to his colleague, who was not impressed. "Oh, wow. Look at your website. It doesn't look very impressive. Are you sure you even know what you're talking about? How old are you?"

"I'm thirty-two," Jacob said, working to keep the edge out of his voice. "We're small and don't have all the bells and whistles. But we got something that really works well with the problem you have."

"Yeah, I'm not sure. We never heard of you guys, and we really got a big problem here. If the big providers in the middle of the exhibition hall can't come up with something, I'm doubting anyone so small as you with limited resources can. It was nice speaking with you. What'd you say your name was?"

"I'm Jacob, with Lucera," he replied, trying to sound as professional as possible to mask his youth and his company's small size.

"Yeah, we'll look more into you later online. But for now we're headed back over to the exhibition center and see if we can find somebody on the floor who can remotely come close to what we are looking for."

Jacob just looked into disbelief. He heard the problem, approached, and presented the possibility of fixing it. Even prospects in dire need still didn't believe in a small firm like Lucera being able to address their problems. Jacob was growing more envious of the big boys whose brand reputations seemed to automatically give them access to clients' confidence. Still, Jacob found it ironic that those same big tech companies, once you got through the door, often couldn't address the main problem after all.

Jacob looked forward to the end of conference and getting back into the office.

Jacob kept fiddling with his tie. He was not accustomed to being dressed up in what he called "the monkey suit." His mom noticed him fidgeting and said, "Jacob, don't worry, baby, you'll get it this time."

"I know mom, I'm just nervous. This new startup award would be a fantastic boost to our brand out there in the marketplace. Lucera could use the recognition."

"No matter what happens when they call that name out, you put everything you could into it," Jacob's father said. "There is a lot of competition. I'm very proud of what you've done here. I've know your solution is innovative, and even if it doesn't go your way tonight, I know they'll be other nights for you".

"Mom, Dad, I appreciate it, but it's getting frustrating to be lost in the noise of the marketplace. I am struggling to get my message in front of prospects. And winning this award would turn that around, especially considering it's a panel of my peers who are evaluating this type of solution technology. They have a keener eye to what actually works out there. I just want a chance for us to be recognized for what we've done."

The event host came back, and Jacob braced himself as the lights came back up on stage.

"The Innovator of the Year Award goes to..."

Jacob couldn't hear anything. But as he saw his parents' faces, he knew it wasn't him. It didn't even matter who actually won. All he knew was once again, he'd been passed up and overlooked, once again held back from getting in front of people who needed his solution.

CHAPTER 5

AN OFFER JACOB COULDN'T REFUSE

IT'S BEEN THIRTY MINUTES SINCE JACOB ARRIVED AT HIS doctor's office for his annual physical. Dr. Johnson was a very congenial man with an engaging personality, and they had a good rapport. Still, it was a frustrating experience.

When Jacob arrived, he was given paperwork to fill out. Even though Jacob had been going to the same doctor's office for the last six years, he always got that clipboard with a pen attached, asking for personal, insurance, and medical history. As Jacob sat down in the reception area, he looked around the room at everyone else with their own clipboards, writing away at feverless pace.

As he returned his forms to the desk, he couldn't help but overhear another conversation, this one between the two ladies sitting in the row of chairs behind his.

"And as I said, every time we tried to input the data, it came back in a different format. We just still can't understand why the app isn't working the way it should. But it just doesn't scale to what we need, and it doesn't translate all the fields properly, especially when the input's in multiple languages."

"Ah, I know," said the other lady. "And it's just like this damned doctor's office. Customers have to answer the same questions over and over, and then when they call in for support, they have to give the same information over again. I wish we could populate all of that context on the front end when a customer is signing up initially for the account."

The first lady agreed. "If only. Our developers just don't have the time to program in all three languages at the same time, and for both platforms. It's very problematic and costly for us to do this. It's causing us a lot of problems."

The second lady clearly felt exasperated. "I know this is something we really need to solve for. We just cannot find any solutions internally, and everything on the market is way too expensive or doesn't have quite what we need. 'Too much customization,' they all say, and we should 'just take that functionality out of the product.' But that's the core function of our business."

Jacob turned around in his seat, marveling at the percentage of Lucera's prospects who entered the pipeline via eavesdropping, and said, "Excuse me, ladies. Sorry to interrupt your conversation. I think I might be able to help."

They exchanged business cards, and he said he would follow up with them to discuss further.

Ten minutes later, the nurse's assistant emerged from the side room, and called his name. She showed Jacob to the exam room, where she proceeded to give him another piece of paper to fill out while she asked him another series of questions. As she took his vitals, including blood pressure, and temperature, she entered the information into the computer in the room.

Ten minutes later, Dr. Johnson came through the door with a big smile and greeted Jacob.

"Hey, Jacob, it's great to see you," Dr. Johnson said.

They chatted for a few minutes, and Dr. Johnson apologized for being late. They were backed up. Again. He immediately got started on the physical examination. As he went through, he took hand notes about some things and typed in other things on the computer to update his record.

"Everything's looking pretty good. Sorry, I have to rush through here. I need to make up for time to get through all these patients today."

"Definitely understand," said Jacob. "I know how it is. Lots of patients and not enough doctors."

He prescribed some basic medication for an issue Jacob was experiencing. Nothing too serious. But he told him that overall he was doing pretty well. They wished each other well and Jacob departed.

As Jacob drove home, he thought about how it was great to see the doc but the overall experience left him wanting. He wondered when technological capabilities would catch up to the needs of healthcare to improve the experience, from the wait times to the manual information capture to the feeling of being rushed.

Stopped at a light, Jacob checked his phone and saw a voicemail from a number he didn't recognize. He hit play, and a man's voice blared through the car speakers. "Hey, Jacob, this is Tom, from Omnify." Jacob couldn't believe it. Omnify leadership wanted to have lunch with him. Tomorrow.

"Once again, Jacob, I'd like to thank you for meeting with Tom and me on such short notice. I know this spot is a little far out, but we just wanted some privacy and have a chance to discuss something important."

It was a beautiful country club setting, but Jacob had been wondering why they were meeting so far out of town. He didn't recognize anybody around him and couldn't recall being this far outside the city for a tech industry meeting.

"Dan, good to meet you as well. I've seen you on TV doing interviews. I was a little surprised about the call," Jacob responded.

"It was great meeting at the tech trade show a couple of weeks ago," Tom interjected. "I was impressed with what Lucera has developed. I went back to my team and discussed what you brought to the market, and you have something we're interested in. We recently stood up a healthcare vertical and made a few other acquisitions to jumpstart our capabilities in this field. We have very strong EHR capabilities integrated into our Synetra platform, but we still have areas we need to build out or expand on to support our healthcare clients. I told our healthcare business leads about TaskCore, and they were excited to learn more."

Jacob was taken aback. "You guys took a closer look at TaskCore? I'm surprised you have interest in a small firm like mine considering your size."

Dan leaned back and laughed. "Jacob, we track everything out there in the market. We make a lot of acquisitions every year to grow and stay competitive, so we're always keen on looking for that next big thing to add to our own product suite. We love startups who are on the cutting edge."

Tom nodded in agreement.

"Yeah, we try not to miss anything. We're very sharp on what we don't do well, and we think TaskCore could fill one of those gaps. That's one of the reasons we wanted to talk to you," Dan said.

Jacob was growing suspicious that he was about to be made an offer that he couldn't refuse. And just then, Dan looked Jacob directly in the eyes and said, "Jacob, we are prepared to make you an offer that you cannot refuse."

Jacob just started laughing, and Dan and Tom looked at each other, obviously wondering what was going on. "Dan, excuse me," Jacob said, "but I was just thinking those exact same words right before you said it." Dan and Tom chuckled, too.

"So, look here, Jacob," said Dan, "we know we have capability gaps in our platform, and we're always looking to get customers what they need. We've been watching your company for the last few months and have been really impressed by what you've done over there. We even had one of our employees go undercover as a prospect to take a closer look at TaskCore. They were quite impressed with what you had rolling. They validated what Tom observed when he met you at the conference," Dan explained, giving Jacob a moment to let it all sink in before he continued. "I'm not going to sugarcoat this, Jacob. We want you, and we want your company. We want to make you an offer. We want to bring you on board the Omnify family and integrate TaskCore into our Synetra platform."

Jacob was speechless. "Wow," he said. "Unbelievable!"

"Jacob," said Tom, leaning in, "we got the size. We got the scale. We have the reputation. We have the brand already established. We've got a full marketing team. We've got a fully integrated sales team that covers the entire country. We've got the technical support team in place. We've got the developers. We've got everything you would need to take your solution to the next level. What we're looking for is an innovative mind to find a problem out there in the marketplace and design the solution. We think we've stumbled upon that with you, Jacob."

Dan reached into his briefcase and pulled out an envelope. "I want you to think about this for the next couple of days. Review it with your family and business partners. I want you to seriously consider joining the Omnify family."

And with that, Dan and Tom presented Jacob with an offer of a lifetime.

Robert and Jacob sat down at a table at the coffee shop with big smiles on both of their faces. It had been way too long, and Jacob needed Robert's advice on the offer from Omnify.

He had already talked with Rhonda and Tim, and he had discussed it with his parents. But he was still conflicted in some ways. Robert had been there with him when he started his own company, and Jacob knew he'd provide great advice at this juncture, too.

From the beginning, Robert was a very supportive mentor to Jacob. He was always there when Jacob needed advice to supplement what his father gave him. Robert was known inside the technology industry as someone who was a great mentor for young entrepreneurs and a great thought leader in the technology field, who just had a good overall grasp on how business works in the real world.

"This is definitely some offer, Jacob," Robert said as he took off his glasses and put the letter down on the table. "They really want you and TaskCore. These are some impressive numbers. And their valuation of your company is much larger than what I would have expected. They're being very aggressive about this. But that's not all that unusual. Big companies with a lot of capital to invest in will make bets, and they don't like to go back and forth haggling on the numbers. This is a serious offer, but I can see by your expression that you're a little hesitant."

Jacob nodded. "Robert, I always appreciate your advice. As you know, over the last couple of months, I've been frustrated that I haven't been able to get traction in the market. I've got a great solution, but nobody knows about it. I really feel that my small size is holding me back, and I don't know how long it would take me to build the brand and scale the marketing and sales efforts that would be necessary to do my solution justice. I know my solution can help a lot of people out there in the marketplace."

"You're right, Jacob," Robert affirmed. "That would be a lot of work. You're probably looking at another two or three years. And as we discussed before, if you wanted to take some venture capital money, you'd be trading in an infusion of capital with the pressures of returning value back to them in a defined timeline. There's a tradeoff. The one thing is, when you invite others in, you're inviting in more decisions makers. I know you love the autonomy that you currently have with TaskCore and the focus on building what you think your clients need. But selling could get you the growth you want with a much cleaner break than having to figure out what to do with the infusion of capital. So, there are pros and cons to all of these options."

Jacob sat there, overwhelmed, taking it all in. "Hey, son," said Robert, "I know you've got a lot on your mind. How are you feeling about it?"

Jacob took a sip of tea, thought about it, and looked around for a second before speaking.

"Robert," he said, looking his mentor in the eyes, "I think I want this. I've got the solution, but my company is just not big enough to get the message out to those who need it. But I worry that selling would be giving up."

Robert nodded, and as they continued talking, he recalled the many challenges young entrepreneurs face running startups, and outlined the limitations of taking a good idea and getting that type of market success. Robert talked Jacob through the pros and the cons of such an offer, but ultimately, he let Jacob know that it was his decision.

"Jacob, follow your heart," he said. "If you think you have got what people need but they can't find you, maybe this is the next chapter in your career." Jacob appreciated that Robert had never been one to tell him exactly what to do but rather provided guidance, allowing him to make his own decisions.

"If you want to jumpstart and get attention for your solution faster," Robert continued, "this is a viable path forward. I could tell you hundreds of stories, Jacob, but this might be something good for you to experience yourself, to take the pressure off some of the other aspects of the business and allow you to focus on the solution while letting Omnify scale the reach beyond anything Lucera could provide. So, Jacob, ultimately, this is going to come down to how you feel in your heart. All I can tell you is only you can answer this question for you. Should you stay independent? Or should join Omnify? Why don't you spend a few days reflecting before making a final decision."

Jacob nodded in agreement. He felt much better. As they wrapped up their time at the coffee shop, Jacob gave Robert a big embrace and thanked him for all his support.

Later that night, Jacob called his father, and they spent an hour and a half discussing everything that was in front of him. "Son, this is your decision," his father said. "I'll support you no matter what."

As Jacob reflected alone in his apartment, looking out onto the big city, Jacob knew it was time to make a change.

A few weeks later, back at the Lucera office, the three childhood friends were reflecting on the past.

"It has been a wild ride, guys," Jacob said. "We gave it a great shot. When we decided to build TaskCore, we knew we had something special."

Rhonda agreed. "When saw what a hindrance manual data import was, and we knew TaskCore could fix it. I'm proud of us."

"I have no regrets." Tim nodded solemnly. "TaskCore made a difference in our customers' workflows and gave employees more high-value face time with their customers. We just found out that it takes more than a great product to be successful in this market."

"We're not businesspeople." Rhonda grimaced, and Tim laughed.

"That certainly became obvious," he said. "It was so good having Robert around, though, Jacob. We got a little better at running a business thanks to him."

Jacob agreed with everything his friends said, and he knew the three of them had reached their limits. "I just wasn't comfortable with the pressures outside investors would have put on us to take TaskCore in directions we were not comfortable with."

"I think it's the right move," Tim said. "And I'm looking forward to leading a product development team at X-Tech."

"And I'm excited to start at AdvancedTech," Rhonda said, grinning. "Advanced wearables never saw me coming. Watch out for what I help build, guys!"

Jacob was glad his friends had landed such great roles so quickly; and that the proceeds from the sale left each of them with a nice sum to stash away for the future. And he was looking forward to joining Omnify, himself. "I will work to keep TaskCore's essence intact over there."

"You're not concerned about losing yourself over there, are you, Jacob?" Tim asked.

"I am a little nervous about giving up all the autonomy of a small firm," Jacob said honestly. "The corporate structure and red tape are a little intimidating, but I think I can take TaskCore to the next level with all their resources."

"Make TaskCore even better over there, Jacob!" Rhonda cheered.

Tim chimed in. "I am glad the rest of our team were able to find new opportunities as well. Everyone of them were top notch talent. All their new companies are lucky to have them."

"I couldn't agree more. It was a pleasure working with them all. Let's all raise a toast to the next chapter," Jacob said.

An hour later, Robert and a lawyer arrived and joined the three partners. They brought the final paperwork for Jacob, Rhonda, and Tim, to sign in order to sell Lucera to Omnify Technologies.

"You three look ready for the next chapter," said Robert, grinning. "I'm excited for you all. I know you'll all do great things."

The lawyer spent the next hour reviewing the paperwork with everyone. Around three o'clock that afternoon, everything was completed.

Lucera was no more.

TaskCore was now the property of Omnify.

And with that, the three childhood friends went on their different paths.

CHAPTER 6

NEW BEGINNINGS

IT WAS A WARM, BUSTLING MONDAY MORNING, AND JACOB WAS navigating highway traffic on his way downtown. This was going to be a big day for him: his first day at Omnify Technologies. Omnify's international headquarters were located right in the heart of downtown, occupying the top four floors of the second tallest office building that hovered over the city horizon. People could see the Omnify logo atop the building from miles away.

As Jacob pulled up to the building's underground garage, he swiped his ID badge on the panel and got a warm greeting from the security guard.

"Good morning, Jacob. Welcome to your first day of work here at Omnify."

Jacob thought that was interesting; the security guard responded like he had already known him for years. He navigated the parking garage to the designated executive area, where he found the spot designated for him, his name on a placard in bold letters.

In the elevator, Jacob closed his eyes to take in the moment, envisioning his arrival at his first day at Omnify. A minute later, the doors opened, and the receptionist greeted him with a big smile.

"Good morning, Jacob. Very happy to have you here. I'll take you around the office and show you where you'll be set up. Dan has you in a corner office. Let me take you over there, first" she said.

"Thank you," Jacob said. "I'm looking forward to my first day. It reminds me of the first day of school; exciting times."

As Jacob followed the receptionist around the large office space, he smiled and nodded at the many employees already bustling around. When she rounded the corner to his office and opened the door to a magnificent, spacious suite, Jacob just beamed. She got him acquainted with his new setup and took him around the office to introduce him to some of his new colleagues. It was a great first morning, and a very promising start.

At ten o'clock Wednesday morning, just two days into his new role, Jacob attended an all-hands meeting with the entire team. Dan took the lead, reviewing the previous quarter, reporting on all the new updates and acquisitions, and reviewing follow-up action with each department head.

Then, Dan introduced Jacob to the team, and everybody welcomed him with a round of applause. It felt good to get this type of recognition, and Jacob was filled with pride as Dan bragged on his innovative mind and the high hopes they had for folding TaskCore into their Synetra platform.

As the meeting wrapped up, many people stopped by to shake Jacob's hand and personally welcome him on board the team. He got the chance to meet people from marketing, sales, support, legal, and human resources. It was a little overwhelming, but he knew he'd eventually get to know everybody by name, so he didn't worry too much.

After the meeting, Jacob went looking for Tom, as he was the product manager for all of Synetra. He wanted to get a head start on understanding the company's plans to integrate his solution into the platform.

"Hey, Tom, just wanted to grab a few minutes," he said. "I know we'll have time to discuss this in more detail later, but I just wanted to get your initial thoughts on the timeline for getting TaskCore integrated."

"Oh, Jacob, great," Tom replied. "My team members are working on getting TaskCore into the schedule ASAP. But don't worry, you'll have plenty of time to get into details. In the meantime, just spend your first week settling in and getting a feel for the team and our solutions. Don't worry, Jacob, we're definitely going to get to TaskCore."

"Thanks, looking forward to that," Jacob responded. He wanted to be a team player, but he wished he could get a few more details, even at this early stage.

The next day, Jacob had a Zoom call with Danielle to discuss some requirements for a new integration point between TaskCore and one of Omnify's third-party platform providers.

"And this is the SKU that you need for this integration point," she said, moving her cursor over her shared screen. "The product comes in four modules, and here's the pricing for it. And here's the level of effort from our team to actually do the coding for this particular feature here. And these are the technical requirements on your side."

"Crystal clear, Danielle. How long do you think this will all take?" Jacob asked.

"We can get this done in plenty of time to meet your deadlines," Danielle assured him. "This is the integration piece you're going to need to provide real-time access to this data source. I think everything is in line here, and based on what TaskCore does and the way it scales, working it all up shouldn't be an issue."

Jacob nodded in agreement. He marveled at the way Danielle rattled off specs, handled the business aspects, and tied it all back to growth implications. Her knowledge of his product and his needs was phenomenal. She was always on her game when he needed help expanding TaskCore's capabilities.

"So, Danielle," he said, scanning through the specs she was showing him, "this piece right here on line five seems to indicate a particular issue we should be aware of."

"Don't worry about that one," she assured him. "We've got that covered, too. It's a different piece of the scope that the services department will take care of, but it will be in place for you when you need it. I know you're on tight deadline to finish this up before the Omnify team completely takes over product ownership."

"No problem. That's good to hear, Danielle," Jacob replied. As they ended the meeting, Jacob was satisfied that he and TaskCore were in good hands.

It had been three weeks since Jacob had started working at Omnify, and he'd continued to receive solicitation from Angela every few days. She knew he was in a new position, and although she'd been convinced before that Lucera could use her company's products, she was even more sure now that he was at Omnify. She was so persistent that Jacob had started to feel bad for her and finally relented to a meeting.

As he settled into his office, he closed the door and logged in to the Zoom session.

"Hey, Jacob," Angela greeted him enthusiastically.

"Hey, Angela. How are you doing today?"

"Jacob, I'm really appreciative of you taking this time," she replied, jumping right into her pitch. "I know you had a lot going on after making the move from Lucera to Omnify. As I said before, I've been following and tracking you for a while, and considering where you are right now, I wanted to let you know about something new we have going on that we think may be useful for you. I just wanted to spend a little bit of time explaining what this product does to see if it would be a good fit."

Jacob leaned back in his chair and said, "What do you have for me today, Angela?"

Jacob did not have the autonomy here at Omnify to make buying decisions that he once had. Regardless, as he listened to Angel pitch this new product, he had a hard time imagining exactly what it was going to do to enhance TaskCore.

He interrupted Angela. "Excuse me, Angela, this is sounding all good, but the fit isn't obvious to me."

He went on to explain the dynamics of the new situation and how, although he was appreciative of Angela's persistence, he just realized he was going down a rabbit hole that wouldn't be a good use of his time right now.

At that point, he thanked Angela for her time and wished her well in finding other prospects for whom this product might be a better fit for. Angela, clearly attempting to conceal her disappointment, thanked him for chatting with her. He hung up and hustled to Dan's office for a meeting he'd been looking forward to for weeks.

When he arrived at Dan's office and introduced himself to Beth, Omnify's chief marketing officer, he knows the excitement on his face was obvious. He was so accustomed to do-it-yourself marketing, with no official staff or budget, and he was looking forward to learning how the big timers did it.

"When I made the offer to Jacob for his company," Dan said to open the meeting, "I knew TaskCore really filled a niche for both the Synetra platform and Omnify portfolio. TaskCore offered something we didn't have yet, that was going to take too long to develop ourselves. Heck, Beth, I don't think we even had the people to develop such an innovative idea. Jacob's product really spoke to our clients' challenges improving frontline productivity to streamline operations. His approach to this problem is something that we just didn't have in-house."

"Thanks, Dan," Jacob responded, pleased to hear this kind of feedback.

"I'm really excited that you and TaskCore are now on board," Dan said, nodding to Jacob before turning his attention to the CMO. "Now Beth, how are we going to integrate TaskCore into our portfolio? I'm looking at what we have in our go-to-market pipeline, and there are a large number of benefits, features, and capabilities here that fit together in a nice, cohesive package. Your team has done a phenomenal job branding and integrating these channel initiatives. So, what are you thinking about how to integrate Jacob's new solution?" Dan asked.

Beth nodded and propped a strategy board up on an easel that usually sat in the corner of the conference room. "You know what, Dan? I don't think we need to do anything particularly special here. I love the product and the solution Jacob has brought us, but our customers know this version of the Omnify brand so well. I don't want to do anything too radical here that makes it look like TaskCore is dramatically changing things. I think we plug it into our core *Productivity Accelerator* pillar, where our customers are used seeing innovations from our newest acquisitions."

Jacob was taken aback by those comments. What was a Productivity Accelerator? But he decided to give her a chance to make her case before he said anything. After all, he was a novice when it came to enterprise-scale marketing.

"And I think we put it over here," she gestured to another section of the board, "and leave it at that."

Dan studied the board and nodded, then said, "You know, I agree. This is not our first rodeo, Beth. We acquire companies all the time, and I don't think this acquisition is any different. Jacob, I believe you are going to be very happy with how his solution mixes in with our other ones. What do you think?"

Jacob was careful to respond in a measured tone. "Excuse me, Beth, but it sounds like you're treating TaskCore just like any other solution. Maybe I am a little too close to it as I created it, but I'm not sure that approach will showcase what it can do."

"Tell us more," Dan prompted.

"If you treat TaskCore like other solutions, you lose what is special about the problem it solves for. The messaging is critical to showing prospects how the product speaks to their pain points. We need to position it as a unique answer to a major problem they have," Jacob responded.

"We have a track record of success with this approach Jacob," Beth replied. "Let me show you some metrics on past acquisitions." She pulled up a PowerPoint deck showing a portfolio of past campaigns and performance metrics from those new solutions.

Dan, sensing Jacob's hesitation, reassured him it was normal not to feel completely confident this early on, especially for those not accustomed to working with experienced marketing teams.

"So," he said, rubbing his hands together, "let's go ahead and discuss the budget, and then we'll get this integrated into the portfolio and blast it out to the market. That way Jacob will see up close how our approach will get us the results we are looking for with TaskCore."

Beth smiled and agreed. "Definitely. Let's get this done."

Jacob was more muted, nodding to avoid making waves this early in the process, he had a strange feeling in the pit of his stomach. He chalked it up to nervousness at being new to the company and acclimating to how they built their marketing campaigns.

CHAPTER 7

MARKETING AND SALES

THE FOLLOWING WEEK, JACOB HAD A RARE OPPORTUNITY to hang out with his dad at his architecture firm.

"This is so great, Dad," he said. "This reminds me of when I was little and you brought me along on *bring your kid to work* day," he said.

Jacob's dad laughed. "I know, son. I'm excited about this project I'm working on, so I wanted to show you what I was talking about at dinner the other night."

Dad led Jacob through the stairs that were still under construction, into a big atrium. "This is the new entrance people will flow through on the grand opening day. I designed the doorways to interact with the flow of the ramp that goes around to the front. If you notice the incline of the windows, a lot of lighting comes in just the right way. So, in the morning, the foyer will be forty percent lit with just natural light. We've taken that into account in the way we're putting in the lighting. It's a new system that will control the lighting for the entire place. The combination of the new LED smart lighting plus the natural lighting will give this atrium a warm and inviting feel."

"An integrated system for such a ubiquitous feature," Jacob said. "Most people don't consciously think about how a building is lit."

"It is the job of the designer to make the *how* invisible while delivering an intentional effect," his dad replied, and Jacob considered how that wasn't all that different from his own work. "And you'll notice here the way we built the in-house seating arrangements. It naturally flows. It doesn't look like seating, but it allows people to get off their feet. There are sensors around the door that will take inventory of people coming in and out instead of having them queue up to be manually checked in. And the HVAC system will automatically adjust to different areas and zones throughout the building."

Jacob looked around, and a series of screens on one wall caught his eye. "Oh, what's that over there?"

"Oh, that is interactive artwork on the wall. So instead of changing out the physical artwork, these are digital displays that are controlled centrally. They'll be able to adjust to the mood of the environment on any particular day," his dad replied.

As they walked around and navigated the construction crew, Jacob was amazed at all the details his dad had put into the architectural plans. Most of the design aspects were both functional and aesthetically pleasing.

When he commented on that, his dad smiled. "We started the design with function first," he said. "Everything from how the doors operate to the size of the walls and the windows to how to support people using the various spaces. Aesthetics are great, but sometimes architects get too caught up in what visually looks good. The focus should be on ease with which people can interact with, use, and experience the space. If you notice here, we have oval-shaped atriums as opposed to square. That helps with the flow of the people coming in and out and gives it a nice breezy feel without backing up foot traffic. Practical yet inviting in a subtle modern way."

In another area of the building, Jacob was curious about the elevator system. It looked slightly different than what he was used to. "This one opens on both sides," his dad said, "and it'll display in real time how long you have to wait before the elevator arrives. It'll take away some of the anxiety people have waiting for those doors to open."

"Oh, I love that, Dad. That's a fantastic feature. I also like the way you blend in the carpeting and the tiling on the floor," Jacob said.

"We wanted a way to shift from the outdoors to the indoors. We don't want people to feel like they're tracking in the dirt from the outside, so we make a transition in between the spaces," his dad said. "Let me take you around up to the top level. You're going to love this."

They spent the rest of the afternoon exploring this beautifully designed new building, and Jacob admired the upmost care his dad had given to every aspect of functionality and ease of use. Jacob was so proud, but even more so impressed with the level of detail that his dad put into how his design was serving people. This blend of functionality and aesthetics was the true art and craft of good design.

Back at the office the next day, Jacob met with Terrence, Omnify's VP of consulting services, in the small conference room on the fourteenth floor. Omnify's large number of products and solutions was of great value to many of its customers. The suite provided a lot of flexibility and modularity for their platform on the broader market. With that, however, came operational and business challenges for Omnify itself. Their business model did not support the ability to have services capabilities for all of those offerings.

In their early years, Omnify could concentrate on supporting the few solutions they did have. They focused on the end-to-end journey, from implementation of the technology all the way down to full customer satisfaction with a broader range of follow-on services. But the cost of keeping in-house consultants and engineers eventually proved too much for their business model.

"Terrence, tell me more about our service partner network," Jacob said.

"Here at Omnify, our approach to services is to get the software up and running as fast as possible and then allow our customers to dive in on their side to make the most use of it," Terrence explained. "It's more cost effective to let our partner network handle implementation. It would be too expensive for us to do it ourselves for all our modules."

"Will our partners go into more depth with clients after the initial technical setup?" Jacob asked.

Terrence nodded. "For some of our core modules like HR-Stack, there are additional maturity services available. But we leave it to our network partners to come up with their own catalogs of service offerings. We tie into that, giving customers a range of offerings depending on the modules they're using. Our customers have come to expect that they'll do a certain amount of the lift on their side, especially when it comes to their internal processes and operational integrations. We've always talked about increasing what we can do in-house, but the more we run the numbers, the more it comes out better to let the partners and clients handle it."

Jacob was skeptical of this "in and out" policy. He tried to keep his face neutral, but Terrence must've seen something flicker across it. "You look a little nervous about this, Jacob," you said.

"Oh, no," he said quickly. "I guess I'm just used to being so hands-on with TaskCore and seeing the implementation through from beginning to end. It's just a newer concept for me to rely on so many other people and for the customers to have to take on so much of the load themselves. As long as the customer is happy in the end, I think we'll be ok."

"Exactly." Terrence grinned and then continued diving into the areas where some of the partners in the network might be able to help out in the case of TaskCore.

Jacob took a sip of his tea and leaned back in his chair, then clicked the link on the onboarding portal. He was excited to dive into an overview of the company and, more specifically, the video product tours of Synetra, Omnify's main solution platform. Especially after that disappointing marketing meeting the other day, Jacob was looking forward to seeing up close how the company had put together the story around their portfolio.

As Jacob scanned the overall picture, one gigantic icon for Synetra stood out. Jacob marveled at the number of modules, add-ons, and additional services that made up all the different solutions in the portfolio. There had been so many acquisitions that happened just over the last three months, including his own company. He was blown away at the speed at which they were able to acquire new companies and seamlessly rebrand and fold the new solutions into their portfolio.

About an hour later, Tom joined him in his office for their scheduled meeting.

"Those videos going through the portfolio were quite long. It took me a while to get through all the different modules," Jacob said.

Although he had noticed that there seemed to be some overlap in certain areas. "I wasn't sure if it was just because I'm still getting used to all the different features or just a function of how the marketing team put together the integrated portfolio, but I did wonder exactly how existing customers digest and prioritize all the new information," Jacob said.

"Yes, it can be overwhelming at first," Tom agreed. Anyway, Jacob was confident, considering how big these guys were and their long history of successful integrations of new products, that wouldn't be an issue.

"I'm really excited about being able to introduce TaskCore, my pride and joy to Ominfy's existing client base," Jacob said. "Considering the amount of time the sales teams have embedded themselves in these existing clients' environments, I don't think there should be any issues introducing something new. That's what really excites me about this opportunity."

"You're absolutely right, Jacob. We have many customers; hundreds and hundreds into the thousands; ready and warmed up to hear about what's next, what's new, what's innovative. I have no doubt they'll be receptive to TaskCore," Tom said.

"That is great to hear," a relived Jacob responded.

"Omnify has expanded into many vertical industries over the years," said Tom. TaskCore will benefit all of them.

"I see healthcare seems to be a big focus area this year," Jacob remarked.

"Yes, we've made several moves in the last year to grow that area of our business," Tom said. "Our healthcare line focuses on catering to the larger providers. We are known for our EHR platform, and we have an impressive mobile OEM-custom tablet that medical professionals love. It integrates all of a hospital's functions into one platform. Still, a common challenge that many providers are facing is the slow coding of services rendered, leading to delayed reimbursements. Every additional day that they wait is another day delayed in account receivables, and they want to improve their turnaround times. Many of our clients' billing and frontline teams have siloed systems. There is a lot of manual data entry and logging on the backend to code the services rendered and then accurately bill providers and clients. We wish our EHR system could be better integrated in that part of the process."

Jacob smiled as he chimed in, "And that's where TaskCore comes in."

Tom nodded. "Now let me walk you through some other aspects of the Synetra platform," he said as he pulled up a new window on his laptop. "Take a look here. This is where the integration points in the platform really shine. As you can see, we built two-way API connectivity between all these modules, features, and capabilities," Tom explained. "In essence, once a client is onboarded into our platform, they're able to up-level into any of these additional modules for additional cost and leverage the different data sources coming from one module to the next. And it's here that we're looking to build that API with your solution. We're really excited about that."

"That's interesting," Jacob remarked. He was optimistic about all the possibilities.

"As you can see, the customer can reuse a lot of the data from one module to the other. We have different places where you can build out the model that will allow the data to flow bi-directionally, really enhancing what the customer can get. This truly is a revolutionary platform, Jacob, and we're really excited about having your solution on board. I think it's going to give us even more attention in this space," Tom said. "Our market research indicated a lot of customer experience angst around this particular pain point, and your solution really hits that nail on the head. So, if they have this module here and that module over there already, and then we can upsell them on yours, and they get a synergy here between the different data sets they've created."

Jacob was really excited about what he was seeing here. The ability to have multiple entry points into getting the conversations rolling with prospects was intriguing. He did have a question, though. "You mentioned the two-way API connectivity. Let's talk a little bit about how that's going to work from a development life cycle perspective. What type of runway do your coders need to build that model out to match what I've already done?" Jacob asked.

"We've got a pretty big deep bench of coders here, Jacob," said Tom. "These guys can integrate anything. If it runs codes, our guys can build an API for it. So based on what you've shown us so far and what they've seen over the acquisition phase, we're pretty confident that with a little rewrite of your code base, we'll be able to get this in."

This last bit raised an alarm bell for Jacob, and he was sure his face betrayed his confusion. "So just a clarification there," he said. "You're suggesting that the code base on my side would need to be changed in some way?"

"That's normal, Jacob," Tom responded in a tone Jacob could tell was meant to be reassuring. "We do this all the time. We normally get in there and rewrite some of the code base to make it work more easily with the rest of what we're doing. There shouldn't be any issues getting everything working with TaskCore. We've got the security, governance, and compatibility teams all lined up to review your code base completely. And there shouldn't be any issues in the way of full integration."

What Tom was saying made sense, but Jacob wasn't convinced. He nodded at Tom to go on anyway.

"But as you can see," he said, "we will need a little bit more work on the user experience. We have an entire UX team that's very good at doing that. Some of the functionalities and features of some of the GUI interfaces on the mobile and web app will have to be adjusted as well. While TaskCore will eventually be available across the platform, we want to focus first on its integration with our healthcare module. That is where the developers will start."

"Oh, interesting," Jacob remarked. "I didn't realize there would be that much change to my code or the user interface. But I trust you guys know what you're doing here. I'm excited about how it will look in the end. It will be good to see the other modules leverage TaskCore's capabilities."

"Yeah, no worries there, Jacob. We'll get back with you as my team continues with the integration plan. We have some deadlines to hit on the healthcare module feature updates. Looking forward to working with you more closely," Tom said to wrap up the meeting.

At the beginning of the next week, Jacob finally had the opportunity to sit down with Beth again to get a better picture of Omnify's marketing efforts. Having learned more and more, he still wasn't sure her lowkey go-to-market plan was right for TaskCore, so he wanted to learn more.

"It's definitely been a whirlwind so far," he admitted. "I'm still getting my sea legs, but I'm loving the interaction with everybody," Jacob said.

"That's great to hear, Jacob. We love having you on board," Beth responded.

"I met with Tom last week to get a tour of the Synetra platform. I'm excited about how TaskCore fits in with the overall portfolio. I'm really interested in figuring out how we're going to get the word out to the existing client base about this new offering. I'm excited to discuss the marketing strategy further," Jacob relayed. "I really think there could be more to it than we discussed previously."

"Jacob, I already talked to Dan and the rest of my team," Beth responded. "We've made quite a bit of progress already and working up a plan to fully integrate TaskCore into the Synetra story. I know you came from a smaller startup, but here at Omnify, we have a full stack marketing department. We've built a go-to-market strategy that's second to none, and we do a great job of rolling the new talent and technology into our own. Every time we acquire a new company, we sell more of everything. The secret sauce is to fully integrate the messaging of the new product capabilities into the larger portfolio story. Our mission is to let everybody in our existing client base, target accounts, and white space, know about how all our new offerings work with what they already know. We think TaskCore is going to bring even more attention to the portfolio value of Synetra."

Jacob nodded agreeably. "I definitely want to talk about some of the things we're going to do to get attention. Obviously, I've struggled in the past with certain marketing channels, such as trade shows and conferences."

"Those are definitely not going to be a problem here," Beth said enthusiastically. "When we go out to the field, we go big. We do national PR, we do TV ads, and we do a lot of social media ads. We go out to all the major tech conferences around the country and overseas. We have a large presence on the ground, from our booths to our collateral. We drive a lot of foot traffic to our booths, but we also host after parties and meetups. We have a lot of branding assets in play. In the past, we had mobile billboards driven around the streets outside the conference venue. And I know you're really interested in the conference we have coming up in the next couple of months out in Las Vegas. Let me pull that up right now so you can get a feel for what we normally do with that and how we're going to integrate TaskCore."

"Oh, that would be fantastic," Jacob said. "I'm looking forward to seeing how some of my prospects walking through the floor are going to learn about TaskCore now that it is under Synetra."

"Yes, we're definitely going to get information on TaskCore to all the customers and prospects even before the conference. They're going to have a heads-up that we have something new going on. We'll also have a couple of breakout sessions to go over what's new in our platform. And then here, we're going to have a spot here where we really highlight what you do with TaskCore." As Beth walked through it all, Jacob got a little lost in some of the lingo that she used. Some of the marketing terminology was unfamiliar to him.

"One second. Back up right there, Beth. What's that?" Jacob inquired.

"Oh, that's just the way we track all the people coming through the booth," she said. "And then on the backend we track the people who have looked at some of our products on the website and integrate all the different touch points. So, it's multi-touch attribution, and our data warehouse on the backend allows us to tailor the messaging to each customer."

"Oh, that's very interesting," Jacob replied. He never could've done something this sophisticated with Lucera. "So, I do have a question for you. Looking at this agenda, it looks like you guys are introducing fifteen new features across seven new product lines inside of Synetra. Wow. That's a lot."

"Yes," said Beth, "we do acquire a lot of new companies, and we have a very aggressive development cycle for all our modules. We develop and roll out a lot of different feature enhancements and changes all the time."

"How do you plan to get my product to really shine in all of that?"

"Oh, no problem, Jacob," she said. "It's a key part of the Omnify portfolio's brand that we are always innovating. So even if they don't directly recognize what yours is doing, many other new products we offer will be highlighted."

Jacob paused for a second and had to let that sink in. It seemed like Beth was going back on the promise that everyone would know about TaskCore even before the conference.

"Beth, are we thinking there's a chance that a lot of our prospects and existing clients might not even notice that TaskCore is available at this trade show?" he asked.

"Well, obviously you're part of a bigger family now, so you're not the only one that has something new to offer. We try to get all of our new products and features out at a holistic level," she explained. "Our main goal with this strategy and this integrated campaign is to make sure our clients know that there's always *something* new. Even if they don't directly realize TaskCore, there's a great chance they'll follow up for more information on other modules."

Jacob pondered that. He understood that he was in a different world now, and TaskCore was not the only solution on offer. However, he was still concerned that, since this solution was very specific to very specific pain points, TaskCore could be lost among all the other new updates the company was broadcasting. But it was not a concern he wanted to linger on. After all, this go-to-market strategy had a proven success rate across the board. He was excited about the possibilities.

Now that Jacob understood the marketing approach a little better, he was curious about the sales process. He asked Gary, VP of sales for Omnify, if he could sit in on the sales leadership team's weekly strategy meeting in the 100-person-capacity conference room.

Gary was legendary in both the organization and the industry. He had over thirty years of experience running sales organizations from mid-size to large. He himself had carried a bag before, a sales term to describe someone who must perform against leadership's expectations to earn a living. Gary had personally knocked on doors looking for people to pitch products to early in his career. So, he knew the ins and outs of sales; the realities and difficulties of converting prospects to paying customers. He'd worked up the ranks when he arrived in Omnify over ten years ago, up to territory manager, regional leader, and eventually his current role.

The sales team was the tip of Omnify's spear out in the marketplace. They were the ones knocking on doors, making cold calls, blasting out email solicitations, connecting over social media, meeting with prospects, and maturing relationships with existing customers, while Gary oversaw the overall strategy and approach. His peers often praised him for designing the blueprint for Omnify's impressive year-over-year growth in the last decade.

In this room were all the sales managers that worked for Gary. They oversaw the individual sales reps that populated all the regional field teams.

As Jacob settled in, Gary brought the meeting to order and gave an overview of the sales numbers from last quarter followed by a projection of the updated sales targets that management had passed down to his team for the remainder of the year. He wanted to make sure everybody was on board with the new numbers. It was almost a ten percent increase in their targets from this time last year. Gary had fought hard for the quota they had been assigned based on last year's performance, but he was very encouraged by the ability of his sales team to go out there and perform.

Gary reinforced Omnify's sales approach, methodology, and unique way of approaching the market. He hammered home the importance of maintaining strong relationships with existing clients and urged everyone to hit the pavement into white space to find new logos to add to their client list.

Gary, himself, was excited about all the new acquisitions Omnify had made, and he directed everyone's attention to Jacob.

"One of the reasons I am confident in this team's ability to hit these updated targets is because of the additions of more upsell opportunities from recent acquisitions. And we're so happy that Jacob's company is now in the Omnify fold. We're very excited to have this sales team and organization go out there and position great innovations like TaskCore to our clients," Gary said. That prompted cheers and applause directed at Jacob, who stood momentarily to acknowledge the praise.

Then, Gary pivoted back to the Synetra portfolio storyline. The platform allowed the sales team to have many different conversations with both prospects and existing clients. Gary hammered home the need for each sales rep to sell one more thing to every customer they served. With that, he assured his team, they could easily hit their numbers.

Jacob listened intently. It seemed like the sales team had a very large catalog of products to go after. Just like with Beth's marketing strategy, he was concerned TaskCore would get lost. So, when Gary opened up the floor for questions, Jacob raised his hand. "How does sales leadership approach their teams in the field about identifying client pain points and challenges versus pitching new modules?" he asked.

Gary went into an explanation of their approach of landing and expanding. "We don't necessarily focus on just one thing or one particular problem but on leveraging our brand recognition and personal relationships to get a foothold and then expand. Our clients are always receptive to hearing about what's new with us. Over time all of Omnify's customers will be educated on TaskCore."

One of the seasoned reps chimed to help answer Jacob's question, too. "Everyone in here is locked into selling the brand and platform. Our teams will shake out enough of our customers' spend to spread around to all of our modules. Jacob, you will see a piece of that shake out your way." He smiled and nodded to others in the room.

With this, Jacob understood that their focus would never be TaskCore specifically, but hitting their individual target numbers with whatever modules they could upsell at the time.

He knew he would have to be patient with this approach.

Later that week, Jacob went on the road with Ben, one of the sales reps from the local field team, to sit in on meetings with some of his existing customers.

"I appreciate you allowing me to come along on these meetings today. I'm getting a good feel for the type of customers who put their trust in Omnify," Jacob said.

"No problem, Jacob," said Ben. "It's good to have someone along for the ride. I usually have my sales engineer with me, but he's out on paternity leave giving all the support he can to his wife and newborn." They were quiet for a moment, until Ben pulled off the highway. "We are almost there," he said. "The client we're about to meet has been with us for over five years. They are very satisfied with our team. We've had great success in landing and expanding, getting them to upgrade and add additional modules every year."

Jacob was curious to see how we would position TaskCore.

But, Ben explained the same concept he'd heard from both Gary and Beth: "It's not necessarily about any one specific new product, but getting a feel for where the client stands at that particular time and identify any unmet needs something from the Synetra platform could address."

When they arrived at their meeting, they took care of general introductions and then got down to business. Ben reviewed what he and the customer had discussed in the past and got feedback on their satisfaction with existing module subscriptions. In addition, Ben inquired to see if there were any new projects the customer was planning. Jacob recognized the standard sales cycle and probing. He was curious and excited to see how Ben would handle introducing the new things the platform could provide.

The conversation drifted into personal pleasantries and talk about the past between Ben's platform updates. Jacob got a bit lost trying to follow the thread, but the whole conversation sounded a little bit like an infomercial, he thought.

Ben went through the long list of new things that they could provide and pressed very strongly for the customer to consider buying at least one. Number nine on the list was TaskCore, but before Ben could get there, the customer interjected. "This is great and all, but Ben, as you know from our last few conversations, budgets are tight right now. We'll probably look at your Robotic Process Automation module for a small pilot program and perhaps increase our licensing on the Enterprise Resource Planning module later this year, but that's about it for now."

"Of course. Totally understood," said Ben. "The main thing was I wanted to catch up with you and see how you guys are doing."

Jacob was fascinated by the fact that Ben didn't seem interested in identifying new pain points, challenges, or opportunities for the customer's organization. The meeting had mainly been an opportunity to show off the new menu. This was both intriguing and a little concerning to Jacob, but he thought, "Hey, this is the way the sales team operates. They've been successful." He was willing to learn their ways.

Still, though, he wished he'd had a chance to explain the benefits of TaskCore further.

The next few meetings went the same way: a roll call of new offerings and no luck getting to talk about TaskCore. Maybe next time, he hoped.

CHAPTER 8

PERSPECTIVES

That evening after work, Jacob pulled into the parking lot of his mom's hospital. Her car was in the shop for the day, so he had volunteered to drop her off and pick her up. When he entered her wing on the fifth floor, the receptionist greeted him and let him know that he would tell her that he was there. After about twenty minutes, his mom came out and they gave each other a big hug.

"Hey darling, how was your day?" his mom asked.

"Fantastic, Mom. How about you?" Jacob replied.

"It's been a long day, Son. I'm looking forward to getting out of here."

They decided to stop off to grab dinner before heading back home. Once they were settled at a table, Jacob inquired about her day.

"It's been a long one, Jacob. A lot of patients to see and not enough time for charting. I'll have to get in a little bit early tomorrow morning to catch up on notes for some of the patient records."

"Oh, that's interesting, Mom. It seems like you have to spend a lot of time doing notes," said Jacob. "That kind of stuff took up a lot of my last visit with Dr. Johnson."

"How's Dr. Johnson doing these days?" his mom asked.

"Oh, very good," he said. "He says to tell you hi."

"I haven't seen Dr. Johnson in quite a while. Is he taking good care of my special boy?" she asked with a grin.

"He is doing a great job," Jacob replied. "I just notice, though, that wait times are getting longer at each appointment. He has to spend a lot of time entering records on his computer, too."

"Yeah, Jacob, when are you guys in technology going to help us out a little bit better?" his mom joked "Your mom could do with an hour or two less of data entry."

They enjoyed catching up on the day for another forty-five minutes before Jacob took her home. He thought about what his mom had said. He wondered how technology could be better leveraged to improve the experience for both the doctor and the patient.

"Robert, it's good catching up with you," Charles, Jacob's dad, said. "It's been too long, my friend."

"I know, I had the same sentiment", Robert replied as he dined with Jacob and his dad one Sunday evening on the riverwalk downtown. Then, Robert turned to Jacob. "How are things at Omnify? Feeling all settled?"

Jacob hesitated before he answered. "They're going well, in general." He paused, hating to admit he was questioning the decision Robert spent so much time helping him make. "Though I worry sometimes that TaskCore will get lost in Synetra's enormous portfolio. It's hard watching it take a backseat to other solutions."

Robert nodded. "Do you think you made a mistake?"

"I don't know," said Jacob truthfully. "I just wish I hadn't had to give up so much control over TaskCore."

Jacob knew Robert and his father recalled the discussions they'd had before he went to Omnify, including about potentially finding venture capital. An infusion of cash to help Lucera scale and become more competitive had been a viable option.

Charles nodded, as if reading his son's mind. "But you know taking external funding has its own downsides. I ran into that problem when I started my firm. VC firms had just started to get into the architecture and construction industries. They didn't understand how we operated as well as they needed to. It caused a lot of problems that took me a while to untangle."

"You handled that situation very well, Charles, and made an incredible recovery." Robert said. "Jacob, don't forget that, once you take money from an outside source, you're no longer in full control. Often, you're captive to generating a return to your investors on their timeframe. I run into that with many of the early-stage startups I mentor. I always tell them of the challenges of giving up part of your ownership and being beholding to an outsider who may not share your exact vision and mission. You would've had to give up your full control either way."

"You got that right," Charles nodded. "You have to focus on your why as you make those decisions; and I know you did, Jacob. Focus on your purpose. If you need to have outside help, that's one thing, but you've got to be careful how that help might influence your future decisions. You may find yourself tracking quarterly returns and lose focus on those you actually serve: your clients who came to you to solve their problem. It's always a fine balancing act."

"But how much control should you give up?" Jacob asked. "Some days I feel like I lost all of it."

Robert replied, "There are pros and cons to staying on your own, getting outside help, or to selling to a larger corporation. For right now, I think you made a great decision to sell Lucera. Hang in there, and see how it goes. And besides, you know I'll keep an eye on you at Omnify."

"I know you will," said Jacob, smiling. "Thanks, Robert."

"Yes, thank you," echoed Charles. "And so will I, Jacob. You're my pride and joy, and I'll always have your back."

Feeling relieved to be reminded of how much these two great men supported him, Jacob turned the conversation away from himself, and the three of them spent the next couple hours just catching up on all the things going on in their respective family lives.

Weeks later, Jacob finally had a chance to catch up with his own dearest friends. As he checked in with the hostess, he heard Rhonda's voice. "Over here, Jacob!" she called. He glanced over to the corner and saw Rhonda and Tim already at a table. Jacob thanked the hostess and went over and gave both old friends big hugs.

"It's great to see you two," Jacob said. "It's been way too long since we last hung out." He took a seat and took a good look at his former business partners. "Hey, Rhonda," he said, "what is that in your hand? Doesn't look like regular glasses."

Rhonda was always tinkering with something.

"Ah, these," she replied with a sly smile on her face. "They're not your standard smart glasses for sure." This is something I've been working on in my spare time, and I couldn't wait to show you two."

"Let me check those out," Tim said. "They look cool!"

"Wait, let me activate them for you." Rhonda fiddled with them a second before placing them over Tim's eyes. "Do you notice anything different?"

"Wow...wait...is that a virtual screen?" Tim marveled out loud.

"Let me check that out!" Jacob swiped the glasses off Tim's face and tried them on for himself, the virtual screen appearing immediately in his line of sight.

"Now, look to the right, then down, then up, and then to the left," directed Rhonda. Every time Jacob moved his head, the projected image in the glasses switched to a different virtual screen. "Pretty cool, huh?" She was beaming. "That is the next level of human-machine interface. Who needs a computer monitor?"

Jacob always imagined that, if the future included a fully functional robot that could mimic humans, Rhonda would be the one that built it. She loved experimenting and putting together parts to make things. He thought that maybe she would create the eRobot that empties the washing machine and puts dishes back in the cabinets.

He joked with her. "Are you sure you don't have a side company you never told us about?"

She smiled. "You know I just like to build things, Jacob. When I get my hands on electronics, I love putting different parts together and seeing what they can do."

"So, we now know what you have been up to lately," Jacob said. "How about you, Tim?"

"Oh, you know me. Always looking to make processes more efficient. I've been working to improve the IDE for prompt engineering integration. That would make it easier for my fellow developers to debug performance issues in their modules," Tim responded.

Rhonda chimed in at this. "So, not only are you the best coder I know, but you're helping your colleagues level up their own games to build better apps faster."

Jacob marveled at his buddy. "You are always on the cutting edge,

Tim." Tim was also a master of infrastructure-as-code. He had an innate feel for merging the management of the infrastructure to the code it ran on.

"How about you, mister visionary problem solver?" Tim asked. "What's new with you at Omnify?"

"Yeah Jacob," Rhonda jumped in. "What problem have you spotted and solved in this restaurant while we have been chatting?" All three of them laughed.

Jacob understood how humans interface with the world. If he saw someone struggling with their computer or phone, he could deconstruct all the components of the physical, technical, and user elements to isolate the design inefficiencies. In school, Jacob's attention would tend to zoom in on problems.

"Well, now that you mention it Rhonda, I didn't like the interface on that tablet the server used to take our order. It's not an efficient flow to show the menu items..."

And ten minutes later, Tim interrupted Jacob's diatribe. "Hey, buddy, let's get back to your company."

"Sorry." Jacob grinned sheepishly. "I've been running into a performance issue with TaskCore that has been pesky to track down. I'm pretty good a coding, but not at your level." He laughed.

"Tell me more," Tim asked, and the three spent the next twenty minutes diving into the details until Jacob had a better feeling for how to direct the developers to fix the issue.

"I wish you both worked at Omnify with me," Jacob said.

"Buddy, you know you can't afford my rates." They all laughed at Tim's faux arrogance.

"Don't worry guys," Rhonda said. "Someday we will all team up again and save the world."

Jacob knew this trio was a dream team for imagination and experimentation. While they couldn't crack the challenges of running a business, if you put the three of them inside of a room and gave them a problem, they would come up with an innovative solution. The strengths they each brought to their work as a group all drove their success in their individual professional paths, too.

CHAPTER 9

BUMPY ROADS

A FEW DAYS LATER, THE DEVELOPERS HAD USED THE ADVICE Jacob got from Tim to clear up the bugs that had been slowing things down, and Jacob took the opportunity to shadow Terrence on a TaskCore implementation while things on the ground were a little less hectic.

"You will need to provide us with the API keys for these three services," Terrence was explaining to AdventCorp, a new customer. "Then my team will validate each one, making sure the authentication and connectivity are good. Next, we'll pull in a sample of the data set from the schema to make sure everything looks accurate. And then we'll run for the next two weeks in QA."

"That sounds good, Terrence," the client said. "Will you be providing some documentation along with that?"

"Yes," Terrence replied. "Our lead services consultant should have sent you a link to a portal, which has all the documentation necessary for your side."

"That sounds good. And then, after we get the sample data in, are you going to help with onboarding all the fields?" the client asked.

"Actually, that will be on your side of the house," Terrence said.

"My team will just take it up to verifying bi-directional connectivity and validate that the data can move."

"Interesting," the client responded. "I'm not exactly sure my team has the expertise to understand all the fields coming in. We'll take a look at the documentation, but if we need any more assistance, is your team available to help?"

"We can get back to you on that if necessary. We can look at a change order to work in some of our consultant's time to help out with that," Terrence responded.

And as this conversation went on, Jacob, who was in the back of the room, grew a little concerned about the potential gap in scope between implementation and onboarding. It didn't seem like Omnify had built in enough time to get this project done to the customer's satisfaction.

"So let's go over the rest of the timeline here," Terrence continued. "We should be wrapping up here shortly."

But Jacob felt like he had to chime in at this point. "How are you guys going to measure the success of TaskCore on your side?" he asked the client. "Do you have specific success metrics your business will be looking at?"

"Not necessarily," the client responded. "We just assumed the Synetra platform's existing metrics would cover TaskCore. Is that not true?"

Jacob felt a pit forming in his stomach at this, but not wanting to step on Terrence's toes, he let his colleague reply.

"Actually," Terrence said, "that's a separate dashboard functionality. But I can get you in touch with your sales team. They would be happy to show you what the additional modules would look like and get that priced out for you. But for right now, we just want to make sure we can verify connectivity between your system and TaskCore."

As the call went on, Jacob took some notes on the side. He wanted to bring all this up later with Terrence.

"I just wanted to do a follow-up on that services working session earlier today with AdventCorp," he told Terrence later that day on a Zoom call. "I had a couple concerns about how the implementation was going."

"Shoot," Terrence replied.

"Normally, when I deployed TaskCore, I probably put in more hours than I charged for to make sure everything looked good. I know we're not looking to undersell ourselves here, but I'm worried the potential change order you mentioned might make the process look a little kludgy. Can you explain a little bit more about how the services side came up with that onboarding scope?"

"Absolutely," Terrence replied. "We want to keep our margins healthy on the services side, so everything after basic connectivity is going to cost extra. That's something we tell the sales team to bring forward when they're trying to sell a solution. Our customers are used to this process right now, though I understand it's not how Lucera approached services. I think we're still figuring out exactly how TaskCore operates here, but we're pretty sure the basic functionality will be enabled."

"So, how are you guys measuring customer satisfaction?" Jacob prodded.

"That's handled by another team. I have to make sure my consultants are being fully utilized and we close out projects on time. But the project team can follow up on satisfaction if needed. I can introduce you to that team if you would like to meet them," Terrence said.

"I'd definitely like to, Terrence," said Jacob. "I'm looking forward to seeing what customer satisfaction looks like after they start running the solution themselves."

"Sounds good Jacob, I have to get to my next call, but I'll connect you to the project team later today." And with that, Terrence left the call.

Jacob closed out the Zoom window, a little concerned about AdventCorp's ability to get the full value out of TaskCore with such limited support from the implementation team. In this new role, he no longer oversaw implementation services or ongoing support; and he wasn't confident his colleagues were handling those areas the way he would have done.

Eight weeks after AdventCorp's initial TaskCore onboarding, Jacob sat in on a call with Sally, the VP of Omnify's customer support group, and the team at AdventCorp. They'd requested a meeting to discuss some issues with TaskCore, and Jacob had spent time beforehand reviewing the open support tickets. As far as he was concerned, there were far too many.

When the meeting began, Jacob was surprised to see that AdventCorp assembled a comprehensive team, including representatives from architecture, engineering, integration, security, and governance. Also on the call were their CIO and CISO. He wondered why they had so many people in attendance.

Sally opened the meeting by running through her checklist of post-onboarding feedback items before she handed the floor to the customer. The CIO chimed in first.

"Hey, Sally. Appreciate you guys setting this up on short notice. We're having some issues working with your support system."

"Oh, okay," Sally replied. "We definitely want to hear about that. We're here to make sure TaskCore is fully integrated into your instance of the Synetra platform. What do you have on your mind?"

"Well, we have a number of support tickets open currently with you guys, and the response hasn't been as good as we expected," the CIO responded. "We've had a couple of other of your modules already in place on the platform, and those seem to be doing okay, although we've had some issues here and there. But for TaskCore in particular, we're not getting the value we thought we'd be paying for when we saw the demo months ago."

"Oh, I'm sorry to hear that," Sally said. "Can you expound on that some more?"

"Yes, my team is struggling to make use of all the data internally, and to actually train the end users on how to use the product properly. Technically, the selected components are in place, but there's some data that's coming in from different sources on our side that we're not exactly sure how to make use of."

Sally dug in deeper over the next hour with the CIO and his team.

Jacob took quite a few notes, trying to get a feel for how Sally was handling the call. He didn't want to step on her toes, either, especially since this was part of the initial batch of implementations Omnify was doing with TaskCore. Jacob believed that he'd done as much as he could to offload his knowledge of his product to the services and support team, and he didn't want to interfere or short-circuit what they were doing.

Jacobs began to regret not having more time to train Sally's team. He had spent so much time developing roadmaps, working on integration guidelines, and contributing to marketing collateral, that the hours had gotten away.

As he listened to the dialogue, he is noticing a pattern that made him doubt whether AdventCorp would get the value that TaskCore was designed to provide. The initial implementation seemed to be a little light from Omnify's side, though he wasn't sure if that was completely a function of the sellers keeping the cost down by limiting the scope or if something else was at play. He suspected it also could be an issue with the services team configuration, and maybe not having enough consultants familiar with TaskCore to do it properly. But he suspected there was something going on here besides just the technical aspects.

He continued to take notes and ponder. He decided to be patient, hoping maybe it was just because this was the first implementation of TaskCore with a client that had this specific combination of Synetra modules already. But he was determined to keep an eye on it. If the problem continued, he wanted to get to the bottom of it.

"So, Tom, you're saying these three functions have to be stripped out of TaskCore? I'm not sure I understand why."

A week after he'd sat in on the support meeting with Sally, Tom had called Jacob in to discuss product integration, and now he was reviewing a list of TaskCore features that would have to be deprecated in order to complete integration with Synetra.

"Because of the way we built the base operating parameters of Synetra," Tom explained, "anytime we add an additional module to our platform, any necessary structural changes must be made to the new product. Doing it the other way around would destabilize our core."

"I understand that," said Jacob, "but why get rid of so much of the new product's functionality?"

"Let me clarify," said Tom. "When we say "product," we mean what we have after we strip it down and modify it to become an actual module that works with the rest of the platform. Some of the functionalities you've built into TaskCore are not compatible with the way we constructed the architecture."

"Tell me more," Jacob prompted, not in the mood to give up more of TaskCore than he already had.

"So if you take a look at the architectural diagram here on the screen, Jacob," Tom obliged, "you'll see the ways we make the calls to bring the data in. We have to scrub it here and then make sure all the fields of the new module, in this case TaskCore, map in a way that is consistent with our platform's root schema. Now we could create a lookup table, which we sometimes do. But that starts to get into some performance issues, so we just like to rewrite those fields completely. And as for some of the third-party calls that TaskCore currently is making, we need to rewrite those as well because we have a central repository for Synetra that takes care of that."

Now Jacob was really frustrated. "Hold up, Tom, this is key to how my team coded TaskCore. These changes are going to throw off the functionality. I'm also concerned some of these changes could lead to compatibility problems down the line. And then the way you're handling the data pipeline and analytics engine seems to be changing as well. It looks like you may be rewriting over half the TaskCore code base."

"It's probably closer to sixty percent," Tom replied, unfazed. "That's because some of the functionality you developed is already handled by our core engine. And some of the other modules that have similar features to TaskCore have a very established customer base. So, we have to make sure that it works together and we don't cannibalize other modules our key customers already love. Sometimes we have to make some sacrifices on the new product to make sure the other ones aren't compromised. As you see here, the core functionality of TaskCore does remain intact."

Jacob was skeptical. "I'm not familiar with the code base of the other modules, necessarily. I haven't had a chance to really dive into some of Synetra's inner workings, but it looks like a lot is being lost here. I'm not exactly sure exactly how these reductions fit in with maintaining TaskCore's value proposition. I'm going to need to take a closer look at the Synetra code base to really understand that better."

Tom pushed back a little bit. "Okay, Jacob. But this platform has a strict governance and security structure around it. You'll have to put a request in through our GRC desk for access."

It seemed to Jacob like Tom was discouraging him from pushing the subject, but he wasn't deterred. "I'll do that, Tom," he said. "And I'll let you know when I'm ready to discuss this further."

That night, true to their promise not to wait so long to catch up anymore, Jacob, Rhonda, and Tim met at a new sushi restaurant that had been getting rave reviews.

"Hey buddy, are you ok? What's going on?" Tim asked Jacob almost immediately. Jacob knew he wasn't looking his best these days, but for Tim to catch on so quick meant he must be looking really down.

"Yeah, man," Rhonda chimed in. "This is supposed to be a fun night. What's going on with you?"

"It's just things at Omnify are coming fast and furious at me. I had a rough meeting today with the platform product lead, and he basically wants to rip up over half of TaskCore's codebase."

"Whoa..." Tim's jaw dropped. "Wow...what's going on there, man?"

"I'm still wrapping my head around it," Jacob replied. "They're using a lot of different operating systems and programming languages I'm not familiar with. In college, and at Lucera, I just had one or two main languages to focus on. But their platform has so many different modules, and they seem to be using every possible programming language imaginable. That requires them to recode large sections of the new products to interface with their core. It's almost like a Franken-platform."

"I thought it was a single platform," said Rhonda, clearly confused.

Jacob nodded. "From the outside, it is. But underneath the hood, there are a lot of moving parts, different languages, and different integrations. Now, with their VP of product development talking about all the things they have to strip out of TaskCore, the soul of what the solution was designed to do could be lost."

"Oh, man, I feel bad for you," Tim replied. "Unfortunately, this is a quite common situation. Rebuilding from scratch is time consuming, so they take shortcuts to get to market as fast as possible with a single customer-facing interface. It's all business pressure all the time. Man, I spent a lot of time with different programming languages. I'm familiar with situations like this. Tell me a little bit more about what's going on."

Rhonda agreed. "Yeah, let's dig into this, Jacob, and see what's going on."

Fifteen minutes later, Jacob shook his head and said, "I'm not sure I will recognize TaskCore when they're done with all the changes.

But enough about me. What about you guys? What's going on in your lives?"

"I hate to throw good news on your bad news," Rhonda said with a sympathetic smile, "but things are going well for me. I'm designing and building out some new ways to interface with voice, motion, and touch. So, I'm pretty excited about that." Over the next twenty minutes, Rhonda showcased a couple of her new "gadgets" straight out of her purse.

Tim, for his part, had been delving more and more into AI. "I've been programming infrastructure-as-code in a whole different way," he said. "And between GenAI and AgenticAI, there's a lot going on."

After Tim broke down what he was working on, Rhonda said, "Hey guys, we spent all this time talking about business, which I love, but are you guys doing okay on the personal side of the house?"

Nobody had much news on that front, but they caught up on what there was to tell. When it came time to say goodbye, both Rhonda and Tim encouraged Jacob to keep his head up.

"There are a lot of growing pains with being acquired," Tim said. "And with a new role, there will definitely be changes. But stick it out there, man. There's probably more good to come out of all this."

Jacob nodded, grateful for his friends' support and perspective.

CHAPTER 10

LOOKING UP

JACOB WAS A REGULAR AT THE LOCAL TECH INNOVATOR MEET-up, a monthly get-together of young entrepreneurs looking to learn more about how to create, fund, and grow startups in a competitive market. He'd been especially looking forward to this month's event, because Robert was slated to give a talk to the young, enthusiastic crowd.

"Welcome, innovators of tomorrow!" he boomed. "You are the future of what the market offers to other businesses and consumers." He looked intently into the eyes of each person in the front row. "You will create something they did not know they needed, believe they needed, or even know was possible. You will take an idea and bring it to life in your home lab. You will iterate variations of your solution, fail many times, and experiment until you come up with something workable. You will go out there and find capital to build a business and take that idea into the global marketplace. You will speak in your own voice and make yourself known to prospects who are looking for the very thing that you created."

Robert had a way of captivating any crowd. He could inspire people who wanted something different, who were coming out of school, or who were ready to branch out on their own and pursue an idea they had percolating in their mind for years, but just didn't know how to get started.

And this is what Robert was imparting to this audience at this meetup: how to take that first step into entrepreneurship, how to find help, and how to persevere through troubling times. Robert went on for thirty minutes, laying out his own personal background, his inspirations, and the story of how he became the successful entrepreneur he is today.

When he opened the floor for questions, a young lady in the third row raised her hand and said, "Robert, I had an idea, but I couldn't convince anyone to invest in me. What would your advice be?"

Robert cleared his throat and said, "Young lady, that is a perfect question. In this town and in this world, there are many folks with the capital you need, and they are hunting for ideal builders, creators, and innovators like you. You have to find the right one, and when you get in front of them, you have to be prepared not only to show your passion for the idea you created, but to describe a viable business model that can deliver a satisfying return for the investor. A great idea that cannot generate a profit big enough to provide the right type of returns will not be enticing to those with capital."

To hammer home the point, Robert made clear that venture capital does not look for products to invest in. Rather, it looks for viable businesses to invest in. "They invest capital in you with the belief that you will return it to them, plus some. They expect you will generate more income than you spend to get your solution to the market. A great idea without a viable business model is not a business worth investing in."

Robert continued to pass on the insights and the wisdom for the next hour and a half. After it ended, Jacob hugged him and thanked him for once again providing such great insights not only to him but so many others. Jacob always enjoyed hearing what his "uncle" had to say, and this time around, the conversation had gotten him thinking: If he were to do it all again, with TaskCore or a different product, the idea of raising capital to scale instead of selling was starting to seem more appealing.

A few days later, Jacob followed Dan and Tom into a steakhouse near work. As Tom made his way to the hostess stand and requested a private table in the back of the restaurant, Jacob wondered yet again what this was all about.

"Jacob, I'm glad you could join us today," Dan said once they got seated at the table.

"Thanks for inviting me out," Jacob replied. "So, can you tell me what's on the agenda?"

"Hang on a minute," said Tom with a grin, clearly enjoying keeping Jacob in suspense. "First thing's first: What's on the specials menu today?"

After they ordered their meals and drinks, Tom finally spoke up. "Dan, can I be the one that lets Jacob know the news?"

"News? What's going on, fellas?" Jacob cut in. "You know you worry a little when your boss takes you out for lunch for no reason."

Dan and Tom both chuckled. "Nothing to worry about, Jacob," Dan reassured him. "This is great news! Tom and I invited you out here to discuss something we've been thinking about for a while. We've seen the work you've been doing since you came aboard; in particular your dedication to the quality of the user experience; and we think you're ready to take on more responsibilities. You've been here for just over six months now. We are in the process of some internal changes that will entail adjusting some team compositions and reorganizing some of the reporting structure around our modules. And just looking at what you're doing with TaskCore, we think you'd be the perfect person to take on a little bit more responsibility and leadership here at Omnify.

"Hmm, sounds intriguing," Jacob said. "What do you have in mind?"

Dan looked at Tom and said, "Tom, go ahead and tell him."

Tom rubbed his hands together excitedly before launching in. "We're looking at the synergies between TaskCore and the rest of our platform, and we've identified three other modules that are a great fit for what TaskCore does. We think you'd be the perfect person to take the product lead role on these other three modules."

"Wow," Jacob exclaimed, but Tom wasn't finished.

"Yes, in addition to TaskCore, you'd also be the product lead for three other modules." Tom handed him a piece of paper featuring a diagram of all of Synetra's modules, three of them circled in red. "These teams will report to you, and you will lead the innovation and product development for these three in addition to TaskCore. I think this would be a great fit for you. You'll oversee more teams, resources, and budget. I think you really could help push us and grow these other modules."

Jacob leaned back in his chair, took a sip of water, and said, "Wow, guys, I'm honored. This is great. This is a great day. I've always had an interest in learning more about these other modules and leading more teams here. I'm really excited about this opportunity.

Dan patted him on the back and said, "We're glad to hear that. Now, let's turn our attention to this food; it looks fantastic. This is a big day for you, Jacob."

CHAPTER 11

CUSTOMER (DIS) SATISFACTION

"NIA AND MICHAEL, THANKS FOR TAKING THE TIME TO MEET with me," Ben said as he prepared to present them with a proposal for some upgrades.

Jacob had been thrilled when some of Lucera's best clients followed him to Omnify, and he was especially happy to maintain his relationship with Nia and Michael. But they were having some problems with this new iteration of TaskCore, so Jacob had decided to join Ben at this meeting at WellThrive's main office.

"Well, actually, we would like to get more clarity on some of our support requests and a better picture of TaskCore's roadmap," said Michael. "As we talked about before Ben, we would like to see better improvement in support response times. And we have some processes we want to improve on our end that we want to make sure Omnify will be able to support."

"Definitely agree there, Michael," said Ben. "That's why I'm here today. I've been talking to both our product and support teams. We think some of the new things we're working on in the upcoming feature releases of the Synetra platform itself, along with using some of our other modules, can really help with what you guys are looking for. I put together a proposal here for a couple of new capabilities I think might really fit."

Jacob cringed. Had Ben even listened to what Michael had just said. He started to clarify, but before he could open his mouth, Nia interjected. "Just to be clear, Ben, we aren't looking to do an upgrade or add additional modules from your portfolio. We're looking to get more clarity on what Omnify is doing to help improve the support and stability of TaskCore."

"Yes, Nia, definitely, definitely," Ben plowed forward. "I think you guys are used to seeing some of the features that we had in the product before we integrated it into Synetra, but we changed up some things to better fit with our overall platform. We removed a handful of features that overlapped with some of the capabilities in other modules. If you'll look at this"—Ben slid a presentation folder over to both of them as Jacob silently begged the sales rep to pivot—"at least three of the things you're looking for are over here. If you upgrade to these two modules here, I think we can give you a great deal, and you'll get back the functionality you need."

"Hold up," Michael said. "As Nia said, Ben, we're not looking to spend more money. We use TaskCore for a very particular reason, and we'd like to get the value out of what we've already paid for so we can expand our client base around town with more sites. But all you guys ever do is keep presenting us with upgrade options as opposed to improvements and clarity on what we already have."

Jacob spoke up at this point, eager to find a way to actually support WellThrive with this stripped-down version of TaskCore. "Let's pivot here, Ben. Nia, Michael, tell us a little more—"

But Ben interrupted. "I can understand your frustration," Ben interrupted, "but we're really positioning Synetra as that central platform, and some of the data sources and features, including the dashboard and reports you're looking for, are already there. It just happens to be in a different module that you guys aren't currently licensed for today."

"Ben, could we give them access to those modules at no cost, given that they paid for features they can no longer access?" Jacob asked. Nia threw him a grateful look.

"Jacob, I know you're still new here, but that's just not how this works," Ben said with a chuckle that made Jacob's fists clench at his sides. "Full Synetra capabilities can only be unlocked with the purchase—"

Nia, clearly and, Jacob thought, understandably frustrated, interrupted Ben again. "I think we're on a different page here, Ben. Let's end this right now so we don't waste any more time. Why don't you regroup with your team and see if we can stay focused on the current issues we defined in our last email to you. Additionally, we will need to know details of Omnify's third-party tablet support for Android and iOS apps."

"Our team thinks it's in your best interest to buy our custom tablet," Ben replied, still undeterred. "It's a more integrated solution, and our development team is on top of all the patches and updates. We have not officially made the decision to end support for the version of TaskCore you are on, but that most likely will be the last version to support the tethering to the tablets you currently use."

"That is very disappointing to hear, Ben," Nia said. "It sounds like Michael and I have some decisions to make." With that, she opened the door and gestured for Ben to leave. He did, finally having the decency to look a little embarrassed as he thanked the WellThrive team for their time.

Jacob stayed behind to see if he could smooth things over.

"I'm so sorry, Nia, Michael," he started. "Can I take a look at what you guys are doing and see if I can find workarounds to fill the gaps with the tech you already have?"

Nia just sighed, but Michael spoke up. "Thanks for the offer, Jacob, but I'm afraid we're getting to the last straw with Omnify."

"You know we love TaskCore, and we loved working with you and Lucera," Nia chimed in, "but this new iteration just isn't working for us, and the never-ending sales pitches are getting old."

"Believe me," said Jacob, "I know. Look, I'll leave you alone to discuss, but don't hesitate to reach out directly to me if you decide to keep working with TaskCore. I'm happy to provide support personally."

They said their goodbyes, and Jacob went outside to catch up with Ben. This turn of events had definitely taken some of the shine off his enthusiasm about the promotion he'd just received.

"I see you're getting kind of fancy here, Jacob." Jacob's mom smiled at him.

He'd invited his parents to dinner at his place to celebrate the big news of his promotion at Omnify.

"Yes, your mom's right. You're bringing out the good silverware this evening," said his dad. "I like this. You're really moving up in the world son."

"Thanks, Dad," Jacob responded. "It feels good to get this promotion. I'm getting more responsibilities over a larger portion of the portfolio, and having this elevated title and more people reporting to me has me thinking I'm in the big times now." Jacob chuckled. "Hence the fancy dishware." He was doing his best not to let his parents see his concern, but of course they noticed the moment he went quiet.

"What's wrong, Jacob?" his mother inquired.

"Oh, it's just that my new responsibilities mean I won't be as hands-on with the actual products. I'll be managing more than doing, and that's definitely something I will have to get used to."

"Yes, son, that's a lot more on your shoulders now," Jacob's father acknowledged. "Going from managing one team to several can mean a lot more people to be responsible for. How are you feeling about that?"

Jacob paused to consider, then said, "I think it should be okay. It'll be different for me, and of course I'll be giving up even more control over TaskCore, which is my pride and joy."

"I know how dear TaskCore is to you. I hope they don't take away all your influence over there," Jacob's mother said.

Jacob laughed. "I'll still be involved, Mom. But yeah, this is a good thing, nothing to worry about. Now let me get the food out of the oven. It should be ready by now. This is a night for celebration!"

But the celebratory feeling didn't last long. A dreary, rainy Friday morning shortly after Jacob learned about his promotion, he walked into his office to find Sally and Terrence waiting for him.

"Hey Terrence, hey Sally," Jacob said. "I assume you're here to go over those CSAT reports? I was reviewing those earlier today. Not good at all."

The direct customer feedback on TaskCore echoed what Jacob had seen in implementation, and what the rest of the team had been hearing through the grapevine for weeks: Customers are not satisfied with TaskCore. We've got across-the-board dissatisfaction, from stability to compatibility to usability. There were countless service tickets following up on issues that weren't resolved during deployment, and customers were struggling to integrate the solution into their own environments, much less make meaningful use of it.

"Yes," Sally said, "Terrence and I are just shaking our heads at how much dissatisfaction clients are having with TaskCore."

"This is what I was worried about with all the modifications that were made to TaskCore during Synetra integration," said Jacob, blowing out a sigh. "There were a lot of compatibility issues that came up during that process that should have been addressed. But even at that, I was not expecting this level of dissatisfaction across the board. I don't even recognize my solution anymore."

Jacob had had multiple conversations with the product team about their approach to the changes. He'd become frustrated by his inability to limit the loss of features that were important to workflow functionality. With his UX and design background, he understood the importance of certain features and seamless functionality that supported the user experience. While Omnify's development team was very capable, the business pressure to optimize for interoperability with other modules put TaskCore's original functionality and stability in jeopardy.

"I've taken my concerns to Tom with little to no success," he continued. "The business priority always took precedence."

Sally didn't look surprised. "We've seen this before with some other modules," she said. "They're stripped down, and then my team gets the brunt of the customer frustration. We can only do so much."

"Same with my team when providing the services," said Terrence. "We're at the mercy of the code quality and features choices. Once we do the initial configuration, we don't have the expertise or time to do more. I feel bad sometimes that we leave our customers in the lurch. Sorry, Sally, that leaves it to your team to firefight."

"Do you have any insights as to where we go from here?" as exasperated Jacob asked. "Our clients look at me as the face of TaskCore. My most loyal champions are now questioning my reputation. There are people I have known for a long time who are wondering whether I sold out for the money".

"We hear you, Jacob," said Terrence. "The next stop might have to be with man at the top."

Jacob knew it might be time to bring his concerns directly to Dan, but he wanted to get all the data in place to make a compelling case for how to fix the underlying problem in a way that aligned to Omnify's business priorities. Dan was a businessman who looked at the big picture, and the portfolio and platform outweighed any individual product or module. That meant interoperability with the core platform would be the top concern.

Jacob sat with Tom and Gary in the small conference room on the sixteenth floor. The two higher-ups wanted to review some concerns around customer privacy that had come up in a recent governance review meeting for a large sales opportunity with a client operating in both the US and Europe.

Gary opened up the conversation. "Jacob, we've hit a potential hiccup on a big upgrade opportunity with one of our established clients." In short, with more customer data being exposed in cyber breaches, there was more pressure on data custodians and brokers to do more to protect and respect the data in their care.

"Yes, we have a problem with the way TaskCore handles customer data," added Tom. "Not allowing customers to opt out of the collection of certain pieces of personal information has posed some challenges. That's going to be a problem here because this customer operates not only inside the United States, but also in Europe, which means we have to deal with GDPR and all of its additional data privacy requirements."

"The way I built TaskCore," said Jacob, "we took care of all the requirements across the different privacy boundaries in the United States by limiting the scope of data access. We didn't have any overseas clients we had to deal with at that point."

"I know," Tom replied. "But now the data we are going to start collecting for TaskCore, which is also available through API to a couple of other modules, is changing the scope for our customers who use Synetra's primary asset database schema."

Jacob nodded. "That's a big implication that could force us to change or reduce the scope of data collection. But if we reduce the scope to avoid being bound by GDPR, that's going to change the type of data we collect. And these particular fields are necessary to power key areas of the TaskCore analysis engine."

"I know," said Tom. "Maybe we can get by with it right now by just removing the fields so we don't have to worry about a customer opt-out option."

"That would be a loss of functionality," Jacob said, shaking his head. "Customers who buy TaskCore wouldn't be able to use it in their overseas operations. That is not a workable approach."

"Creating an opt-out for the customers in Europe would be a lot of extra work and coding on the Synetra platform itself. It wouldn't be cheap," said Tom. "We've always dealt with this by reducing what was coming in or letting the customer handle the data in-house."

Jacob couldn't get behind that approach. "That would be a lot of extra work the customers would have to do themselves to reformat the data, rewriting code and re-architecting to make it all work properly."

"I know, but our governance team is looking at it and they're not too happy about the implications here," Tom said. "We expect the GDPR regulatory body to be sending us legal notice on this soon. Obviously, TaskCore is in production already and we have several customers using it as is. Without the ability to handle regulatory and privacy rights, we have a big problem for international customers."

Jacob's brain was whirring. How could he and his team fix this?

"As for this customer we're dealing with right now," Tom continued, "we will need a workaround. Let me get with our developers and come up with a plan. Gary, I know you are looking for an update as soon as possible."

Gary nodded. "My team is looking to get this closed before end of quarter. This privacy issue could derail this upgrade opportunity and cause us to miss our targets. I will let our governance team know you are on top of this."

"Guys, I'll get right on it and come up with something shortly with the team," Jacob promised.

"Yes, Lauren, I've got Jacob in front of me now," said Dan into the phone. "I'll definitely go over this report with him." Lauren was Omnify's chief financial officer. She was laser focused on the bottom line and ensuring the financial health of the company. He hung up and spun his chair around, gesturing Jacob to come closer.

"Hey, Jacob, take a look at my screen here. This is what I was discussing with Lauren. These are the latest financials on the platform upsells wins from the TaskCore module. They're not looking good right now. We were projecting a much higher growth rate across the Synetra platform as a whole, and we were counting on some pull-through from add-ons such as yours to get an increase year over year. Instead, we're seeing decreases in both. Not the kind of performance we wanted to see closing out this quarter."

Jacob nodded. "I've been worried about this as well. The more TaskCore gets integrated into Synetra, the more it seems to be taking on a different life than I designed it for. I almost can't recognize my own solution, and frankly, I understand some of the frustration new clients are having with the experience. I've been discussing the implementation approach with Terrence, and of course Sally and I have been reflecting on those recent CSAT scores."

"Be that as it may," Dan said, "we really like our newly acquired companies to scale and perform better than what we are seeing from TaskCore. This is something we're going to have to address to turn around quickly, Jacob."

Jacob paused for a moment to collect his thoughts before responding. "Look, Dan, this is a systemic issue," he said. "The value of TaskCore that it solves for widespread challenges around knowledge intake and sharing. It's got to meet certain design requirements to function at its highest level, but those requirements are being loosened under Synetra. From product development to sales, the focus is gone."

Dan pushed back. "Now, Jacob, you knew upfront what you were getting yourself into when you came on board; and certainly when you accepted your promotion. My expectations were that you were ready to make this work, but it sounds like you're struggling to make workarounds that drive sales."

Jacob knew this was true, but he didn't have time to oversee every step; he needed the team to get on board with his vision. "With my increased responsibilities with the other modules, I have not been able to dedicate myself to TaskCore," he replied.

Dan was unmoved. "That's the nature of climbing the ranks in an enterprise solutions organization. At the end of the day, it is my expectation that you will find ways to right the ship inside the constraints of maintaining a platform with competing demands."

Sensing that he was fighting a losing battle, Jacob decided to take a problem-solving approach and push forward. "I understand where you are coming from. I will make it work, Dan."

With that, Jacob spent the next eight weeks pushing aggressive changes to the current codebase to try to right the ship. He sensed that he was going too fast, but he felt cornered. He was committed to improving the quality of the product and turning around the sales numbers at any cost.

In the meantime, he also wanted to figure out how to better communicate TaskCore's value from a sales perspective, so he asked Ben if he could sit in on a few more sales meetings. During this particular one, Jacob sat back and watched as Ben and his sales engineer sat across from the prospective client's team. Their CIO, CISO, and CFO were all engaged in the discussion, but Jacob had a weird feeling in the pit of his stomach that something was off.

On the surface, the presentation was competent, and the information was accurate, but the whole thing felt rushed, and there seemed to be a missing connection between the two parties.

The sales team explained the features as well as they could be expected to with the training they'd had, but Jacob sensed the prospects weren't vibing with the benefits. They were not convinced that buying this solution would yield them tangible results.

"Ben, why can't we get this functionality from one of your competitors?" the CIO asked. "It looks similar to what we've seen elsewhere."

Jacob bristled at the question. TaskCore's functionality as he'd designed it *wasn't* available anywhere else. Ben appeared undeterred. "We have an integrated approach. Our proprietary tablet integrates TaskCore with your process automation, business intelligence platform, and billing system. You just won't find that kind of interconnectivity elsewhere."

Their CFO shook his head. "That sounds all good, but the price tag on this offering seems very high. Even with a multi-year discount, we aren't sure we can operationalize all these features fast enough to justify the upfront costs."

"We do these types of deals all the time," Ben responded, "and you're not the first to ask that question. I have a financial model we use to analyze clients' return on investments over time." Ben walked the team through the model, and Jacob could see that the CFO was getting more comfortable with the numbers.

Their CISO chimed in next. "I would like more detail on how you secure the data flowing between the apps in your backend platform. You guys have a large portfolio of suites doing a lot of work across a lot of areas. I'm concerned about the level of access we'd need to grant to privileged services accounts and APIs to make all this work."

Ben turned it over to his sales engineer to address the security concerns. And he did, effortlessly, then they both continued fielding questions for a good forty minutes. The prospect was doing thorough due diligence, but the conversation still felt disconnected from the very specific pain points TaskCore could solve.

He interjected near the end of the meeting. "I want everyone to remember why TaskCore was built: to solve real-world business problems. The question you have to ask is what it costs *not* to address your challenges." Jacob took the prospect on a journey through an experience their frontline workers go through daily, contrasting their current processes with a future that included TaskCore."

Their executive team seemed to appreciate how Jacob framed the opportunity, but then Ben interjected again. "And if you guys sign by the end of the week, I can get you an additional five percent off!"

Jacob had to bite his tongue. Ben had managed to take the value proposition out of the air and put it back on the crowded field of costs and features. He thought back to the email that Gary had sent out a few weeks before, encouraging the sales team to be more aggressive in closing deals before the end of quarter to meet sales targets. That was the source of the problem. Ben was following orders from the top. He'd have to talk with Gary about how the team could balance that aggression to close deals with a real focus on value.

CHAPTER 12

BAD TO WORSE, AND WORSE AGAIN

A FEW WEEKS LATER, THINGS WERE STARTING TO LOOK UP. Jacob was finishing up a meeting with Beth to review the campaign promoting new TaskCore features and capabilities, including integration with Synetra's business intelligence module and an expanded AI workflow recommender, plus new support for additional Bluetooth devices. Initial pilots were positive, and Jacob was happy with all the hard work Tom's developers had done in a short amount of time.

Jacob had also joined Ben on a call to review the enhancements with Nia and Michael, and they were excited to see Jacob in a more hands-on role.

Jacob headed back to his office, feeling good about the new direction TaskCore seemed to be taking. But when he opened his office door, he could hear his phone buzzing. He picked it up and saw several missed calls and texts. The latest said only, "Emergency."

His heart started to race as he checked his voicemail messages. "Hey Jacob, this is Jackie. We're having a problem with Synetra. It's too early to confirm, but we think we may have been breached. I have assumed the incident response lead role."

Shocked, Jacob headed down to the Ops Center right away. As he entered, he saw Jackie divvying up different tasks as people poured into the room.

When she saw him, she paused to debrief. It appeared that there had been unauthorized root access into the Synetra platform. "Based on the initial reports from our security operations center, it appears that someone may have compromised one of the API calls coming in through a portal that we traced back to TaskCore," Jackie explained.

Jacob looked at her screen, confused. "That doesn't look like the code linked from my API. I'm not sure how that's working. But we spent years reviewing the code for TaskCore, so some of the recent changes must have left us exposed."

As everybody worked frantically on their screens and phones, Jacob just shook his head. He had never seen this type of issue before. But Synetra's developers had made so many changes to his solution that he himself couldn't keep up with their lightning-fast development sprints. This was yet another headache that Jacob hadn't seen coming.

Jacob scrolled through the rest of his missed callas and saw several from Nia. Of course. WellThrive stored personal identifiable health information in TaskCore, and Nia was understandably concerned about her clients' information being exposed. Jacob sent Ben a quick text update to pass on to Nia while he focused on the response efforts.

An hour later, Jackie updated the team on what they had found so far. The credentials for the API in question had been compromised. Synetra had an automated secrets management system to manage credentials and key rotations, but it appeared that this API key may have bypassed normal process and been hard coded. Something like this should have been caught in the code reviews, dynamic security scans, or one of the pre-deployment pen tests.

How had it been missed? There were too many unanswered questions at this point, and Jacob could really use a sounding board. He called Tim.

"Hey, man, got a minute?" he asked when his friend picked up. "I could really use your help."

"No problem," Tim replied. "Just picking up a little coffee here before I head back to the office. What's up?"

"You may have already heard: Omnify is responding to a cyber breach incident right now."

"I saw that news update," Tim confirmed. "I wanted to call but knew you be buried with the response."

"I could just use a little sanity check here," Jacob went on. "Between me and you, it looks like one of the API calls from TaskCore is pulling information it shouldn't have been and was exposed externally."

"Oh, man, that's tough," Tim said.

"This was one of my concerns about the way the TaskCore codebase was being rewritten to integrate with Synetra." Jacob was ready to launch into a full vent, but Tim cut him off.

"I know, man. It's tough to integrate a solution like that. But let's focus on the breach. Now, you'll definitely want to make sure your responders look at how the API calls are passing through the API security gateway. That's going to be critical. They need to validate if there are access restrictions of the source calls. Also, figure out how many other Synetra modules had access to that same API. Then, look into how credentials are being managed. They usually rotate for those APIs, but that requires coordination upfront between the different teams."

"That's one of the problems," Jacob replied. "It looks like the credentials may have been hard-coded."

"Yep, that would be a problem," Tim confirmed.

"Okay, what else?" As Tim rattled off a list of things to look at, Jacob jotted notes on his phone as fast as he could. About fifteen minutes later, he sighed. "Thanks, Tim, this is good. Now I can make sure I've got my ducks in a row here as I get more feedback from the incident response team lead. I appreciate it, man."

"No problem, buddy. Let me know how things go," Tim responded.

Two weeks later, once the response team had gotten a handle on the incident and the situation calmed down enough that Jacob felt like he could take a breath, he called Danielle and asked if she had time to catch up. He needed a break from the office.

There would be many lessons learned from this incident. And no matter the fallout to Omnify as a result of the breach, he was still responsible for improving the sales numbers.

She happened to have a sales meeting on her schedule. He'd been looking for an opportunity to learn how to improve the TaskCore sales process, and he could think of no better teacher than Danielle.

"It's nice to get out of the stress and stuffiness of headquarters," Jacob said to her in the car on the way to the meeting. "I appreciate you letting me tag along."

"No problem," Danielle replied with a big smile. "I know for the last couple of weeks you guys were dealing with the aftermath of that big cyber breach. I know how time consuming and frustrating that can be."

Danielle pulled up to the coffee shop, and they both went in to meet the client. “There he is,” Danielle gestured to the corner table, where a man in his thirties waited with a laptop.

After greetings and small talk, they got down to business. The client was looking for help integrating several of the systems his company needed for the next phase of their business innovation project to grow their revenue.

Danielle reiterated and captured their current business state first, taking notes on her notepad about the main drivers, the various stakeholders, and the financial goals. She led the conversation expertly, drawing out the client on particular issues and milestones, but Jacob noticed that she did it all in business lingo.

It wasn’t until she had the business state fully in hand that she turned their attention to the technical aspects, first documenting what their current infrastructure and capabilities were and then reiterating what she had heard before in previous sessions with them.

“It looks like you have a lot on your plate to get this initiative off the ground,” Jacob chimed in.

The client nodded. “Yes, definitely. And we need help with this integration work. We don’t have the staff and time to do it ourselves.”

Danielle jumped back in, and Jacob was pleased to have teed her up so nicely; not that she’d needed him to. “That’s why we’re here,” she said with a grin. “So, we’ve already got the proposal built out with timelines and milestones that we can discuss. And then what you’ll see here is value-based pricing. So, we will get the business outcome that you need based on the performance metrics you defined.”

Through the next hour, Danielle took the client through the proposal, outlining each phase of the project step by step and clarifying what his team and her team were each responsible for. The accountabilities were all mapped out, the benefits for each phase were defined, and there was a clear definition of what success looked like.

Jacob was blown away with the efficiency with which Danielle conducted herself. Her company was definitely lucky to have her on board. He didn't know if it was their sales training or her innate skills, but she was more than just a regular salesperson. She was a true business consultant and partner. Jacob was impressed.

He was enjoying this masterclass when his phone buzzed with a text message: "Synetra Outage!!!" Aghast, he promptly excused himself.

Danielle had driven him to the meeting, so he called a ride share to get back to the office. On the way there, Nia called and described the scene at the downtown WellThrive clinic. No one could login or access TaskCore. Not on their computers or tablets. The lobby was backing up with members, and Michael was trying to troubleshoot the front desk system.

Nia had been right in the middle of updating a member record when her system froze up, and one of the clinicians had suddenly become unable to enter any more information into a meal plan. She'd restarted her tablet with no luck.

As the car pulled up to the office, Jacob apologized to Nia and told her he'd update her as soon as he learned what was going on.

"If it isn't one thing, it's another," Terrence belted across the room at Jacob as he walked in the door. "First a cyber breach, and now this!"

"Catch me up," said Jacob. "What's the range and scope of the outage at the moment?"

Sally jumped in. "At the moment, it looks like up to twenty percent of all our customers are down." This was not good at all.

"We're trying to diagnose the source of the problem," Terrence said. "At the moment, our engineers are pointing to TaskCore. There's something with the way it's sending data back and forth across our backplane."

"That's weird," Jacob replied. "With Synetra's architecture, that shouldn't be possible."

Sally shrugged. "Nevertheless, we do have a widespread problem."

At that moment, Dan blew in. "Guys, guys, what's going on here? I got my phone ringing off the hook. Who has a handle on this?"

They spent the next forty-five minutes troubleshooting as a team. It looked more and more like the cause sat with TaskCore. Between the refactoring and the changes made in response to the recent breach, the module should have been more secure. Whatever TaskCore was doing, it should never have been able to affect the entire platform like this. Something serious was wrong.

Luckily it had happened near the end of the US workday hours, giving them a little time to figure things out before clients would need access again. Dan provided updates publicly throughout the night while the rest of the team worked.

The issue was with a function to scale back-end capacity with containers to handle more AI responder requests. The API calls for requests exceeded current loads on the AWS tenant, spinning up more containers in another region. The middleware collector, however, had maxed out its spool capacity to analyze requests for their data warehouse needs. Too many requests had crashed the load balancer, setting off a cascading wave of out-of-control container spin ups that were calling for new licensing for worker nodes.

The dev team was able to modify the code logic to abate overload temporality while they manually brought up more nodes on other regions and from Azure resources.

The worst of the outage lasted five hours, and all services to all

Omnify customers were fully restored by 4:00am, and an exhausted Jacob got home around 6:30am.

A few days later, Jacob needed to see a friendly face, so he found a few minutes in between after-action meetings to video chat with Rhonda.

"Wow, Jacob, you definitely look stressed. That outage is really draining you. I can see it," Rhonda said, trying to console her friend. It's not just that, Jacob said. It's been a tough couple of weeks, between the breach and now the outage. And on top of that, I'm pretty sure we're about to lose Nia and Michael as clients. It's like everything is coming down on me all at the same time."

"What's been going on with WellThrive?" Rhonda asked. So Jacob caught her up on their frustrations with the proprietary tablet and Ben's constant upsell efforts. "I feel for Nia," he said. "WellThrive runs on tight margins already. They pay for a solution that's not working well, and now they're hearing they have to upgrade hardware? Not to mention worrying about their members' PHI in the data breach and then losing access altogether in the outage."

"Wow," Rhonda said. "Talk about a triple whammy. I feel for those two."

"I know. Sometimes I just want to jump in there and say, hey, let's just keep supporting what we've already sold our customers and stop messing with it." He put his head in his hands. "It's all so frustrating, Rhonda."

"I know," she replied, and then she treated him to a patented Rhonda pep talk that both included great advice *and* made him feel a little better.

And yet, two weeks later, Jacob was in the conference room waiting for Nia and Michael. They'd called the meeting, and Jacob suspected its purpose. He replayed all the conversations he'd had with them as their problems with TaskCore had mounted over the last couple of months. He'd tried to help as much as he could, supporting Sally and Terrence on the back end and doing as much as he could with the support technicians, but he just couldn't be as dedicated as he used to be.

After a knock, the door opened and the receptionist brought Nia and Michael in. Jacob greeted both of them, and Nia got right into it.

"Jacob, you know why we are here. This is not an easy decision, but Michael and I need to move in a different direction." Jacob nodded. He'd known what was coming.

Michael spoke up. "A year and a half ago when we met you, we were all in on TaskCore. It provided everything we needed. But after the sale to Omnify, as our needs evolved, TaskCore just couldn't keep up. And as we've discussed many times, the support team's efforts don't match what we were looking for. We're having to troubleshoot on our own more and more, because we just can't get your team to pay attention."

Nia chimed in. "And Jacob, it was more than just technology challenges. Without direct access to you for help with the operational side of implementation, there was always a disconnect. The vision you sold us with Lucera for how TaskCore would interact day to day with real people was just gone."

Jacob understood everything Nia and Michael were saying, and he couldn't disagree. "I understand your frustration," he said. "You're right. Omnify has the technical know-how to maintain the back end of TaskCore, but not necessarily the front end, where people work. That's a different type of skill set that I've honed for many years but Omnify has never invested in. I'm not making any excuses. You're right. TaskCore is not what it used to be, and I'm really sorry for how this has played out."

They spoke for another thirty minutes or so and then wrapped up. Nia expressed her regret of having to make this big decision, and both of them assured Jacob that they had all the respect in the world for him and for his vision, and that if circumstances were to change in the future, they'd welcome an opportunity to work together again.

CHAPTER 13

NEW IDEAS

A MONTH AFTER THAT HORRIBLE MEETING, JACOB TAGGED along as Robert's guest to a breakfast roundtable for marketing leaders, where Robert was giving a presentation.

"Let's explore the perspective of the buyer-seller experience," Robert boomed. "Why do marketers do what they do? In the technology industry, the marketing team is insistent on the prospect knowing that their solution is SAAS-based...has a single pane of glass...their data centers are geographically dispersed all over the globe...they've been in business for fifty years. Why? Why do they feel these details are important early in the sales process? Why does the prospect care about these factoids at this moment?"

"Marketing teams also are very insistent in their literature, pamphlets, commercials, ads, and websites. They're intent on making sure everyone knows their solution is AI-powered. For some reason, they believe the prospect finds this critical. A company spends a lot of time talking about their brand, about how smart their people are, and how many years of experience they have. But why does their prospect or potential future buyer care about any of this?"

Jacob shook his head ruefully. He'd been asking the same questions about Omnify's sales and marketing teams.

"They want us to know that the organization has ethics, that their workforce is diverse. That's really just a basic level of courtesy, accountability, and transparency. So why spend so much time going on about it?" Robert continued. "Those other facts are just table stakes. Prospects won't care in general until later in the sales process when they are close to buying. What they care about is if you are truly providing a solution to a problem they have."

Jacob thought about a joke his mom had always made about car maintenance: "I don't care about *how* it runs," she'd say. "I just care *that* it runs." The same applied to sales, according to what Robert was saying, and Jacob was gratified to make that connection so clearly.

"If one was asking a happy client what's so important to them about your solution, would they mention the technical specs or the company stats? Or, would they talk about how happy they are that their pain, their problem, their challenge has been taken care of because of your solution?"

Robert went on to talk about focusing the sales and marketing messages on conveying the ability of the solution to solve the prospects' problems. "People want to know why they should take time to listen to you," he said. "Innovators have to lead with the problems they're solving, or else nobody will care about the innovations. A good marketing strategy lets prospects know you understand their problem and you can help them solve it."

After the roundtable was over, Robert took Jacob to a three-story office complex.

"What is this place?" Jacob asked, but Robert just motioned for Jacob to follow him inside.

In the lobby, Robert pointed out the kiosks spread around the floor, each with a person or two in front of them. In the center, hanging from the ceiling, was a sign identifying the company as FutureWay.

The walls were covered in large murals showing various family journeys and stages of life: a couple buying a home, parents paying for a child's college tuition, an elderly couple on vacation, and another a young woman starting a business.

"FutureWay?" asked Jacob. "I haven't heard of them. But I like the ambience. What do they do?"

"Let's observe what's going on," Robert replied obliquely.

The two men watched quietly for several minutes. People would come up to the kiosks and interact with the systems, then a few minutes later, someone in a FutureWay uniform would come out and greet them, leading them to one of the various side doors.

Soon, one of the managers came out to greet them.

"How are doing, Carl?" Robert asked. "This is Jacob, the young man I wanted you to meet."

"Good to see you, Robert. Great to meet you, Jacob," Carl responded, shaking both their hands. "Robert asked me if I could show you around."

"What's going on at the kiosks?" Jacob inquired.

"Come, let's check one out," he said as he led them over to one. "Our customers come in looking for white-glove service around all things financial, from school savings to business loans to estate planning. One couple you witnessed a few minutes ago was coming in to plan a family vacation. They interacted with our system to create a visit profile that would prep our advisors in the back." He gestured to one of the side doors. "Once the team in the back gets the notice for what a customer is looking for that day, they run some financial analysis and prepare some goal planning roadmaps, then bring the customer in to talk."

"That's an interesting workflow," Jacob said. He was intrigued, but he wasn't sure he was fulling grasping the model yet.

Carl smiled. "I knew you would find this interesting. We want to immerse our clients in potential outcomes based on their desired financial goals. The more data we can populate and analyze before they even sit down, the more effectively we can do that. The profile they fill out on the kiosk gives us all the information we need to tailor our guidance specifically to their needs." He gestured to a young couple who'd just emerged from a side door, smiling. "In their case, we modeled what their future home could look like from a builder's standpoint, and we matched that vision with the loan process and plan to save for their deposit."

Jacob took out a notepad and pen to jot down a couple of thoughts. FutureWay took common but disparate financial needs and integrated them under one roof, leveraging common backend infrastructure to power a redesigned experience of planning for desired financial futures. About an hour later, they thanked Carl and left to have a seat on the patio to chat about what they had just seen.

As Jacob tried to collect his thoughts, he was interrupted by a call from an unfamiliar number. He hit the "reject" button, but whoever it was called right back.

"I'm sorry, Robert, let me take this," Jacob said, the double-call making him worry that it was something important. He held his breath hoping it was not another issue with Synetra.

But as soon as he answered the phone, he realized it was anything but. On the other end, a very young voice said, "Sir, do you have a few minutes to chat? I have something I know could help simplify your life."

Really? Jacob couldn't hide his exasperation as he said, "No, I don't. Stop calling me, please. I have no interest at all." And he hung up. Jacob knew this young sales rep was making hundreds of calls a day hoping to get lucky and hit their performance metrics. But still, why did they all have to be so persistent?

With that, he had an *aha* moment. What if this kid was from *my* company, trying to sell *my* solution? Were our prospects as annoyed as he was? Jacob wanted to dig in deeper. He thought he may have come across a turning point. But for now, back to his time with Robert.

Clearing his thoughts, he composed himself and got back in the moment. "Robert, that was just amazing. My head is spinning," he said as he looked down at his feverishly jotting notes.

Robert smiled. "I thought you might enjoy that. FutureWay is one of the new investments in my portfolio. They are transforming user experiences. I know you've been going through a lot, and I thought immersing yourself in what they were doing would brighten you up."

"I definitely needed this," Jacob said. "With all the fires I've been putting out lately, I feel like I'm so far away from imagining new experiences and relieving people's pain. I seem to do more managerial and strategic planning meetings along with crisis resolution more than anything else. So, seeing the buyer-seller experience up close; and the smiles on the customers' faces as they got what they needed, was a good reminder of what I really love."

Jacob took a deep breath. What he was about to say felt like admitting failure to Robert, but he had to shake something loose. "Robert, as you know, I've had a lot of different challenges over the last year, and it's just been overwhelming. I've been feeling lost at Omnify."

"I know," Robert replied. "I've seen the decline in your joy, Jacob. You're caught inside the machine. You need a way to reignite your passion for what you were doing."

"I know, but I'm just feeling so trapped," Jacob said. "TaskCore is no longer what it used to be, but I have no time or power to fix it. It's just been heartbreaking to see. I can't recognize it anymore. And I feel so isolated from meeting the customer where they are and understanding what they need. I'm in so many executive meetings that I'm rarely getting out on the chance to actually meet prospects and clients out in the field. I'm having to sneak out into the field when I can."

Robert nodded. "Something else that might be useful for you is a fresh perspective on how great brands are created. There is someone I want you to meet as soon as possible. Lisa is a savant with brand development and marketing."

Robert explained that Lisa was a brilliant branding expert. She was known for turning around struggling go-to-market strategies for companies to ensure the company's value proposition shined and cut through the noise, delivering the right message to the right audience at the right time. Robert often sent the businesses in his portfolio to Lisa to right the ship on their marketing and branding efforts.

"And," he said, "I have to go tend to some other matters, but if you're up for continuing your business day out, I happen to know she's available for coffee this afternoon."

Ecstatic, Jacob thanked Robert profusely, giving his mentor a hug before they parted ways; Robert back to his home office and Jacob downtown for his surprise coffee meeting.

When Jacob got to the address Robert had given him, Lisa was waiting for him in front of the central hub for some of the largest tech companies in the world. They greet each other and then started walking down the street.

"Jacob, what do you think about when you see that logo?" she asked as she pointed to the sign atop the tallest office building downtown.

"That is Hathagan," Jacob replied. "They're legendary. Everyone knows them. Their products are cutting-edge."

"That is an interesting description. Why do you say 'everyone'?" Lisa asked.

"When someone thinks of innovation or building the impossible, they think of Hathagan," Jacob said. "They've had tremendous growth from the classic one-man garage startup to an empire with global reach."

Lisa led Jacob down the street and around the corner and pointed to the top of another building. "How about that logo?" He didn't recognize that one.

"They produce some of the top firmware and embedded chips in the industry," said Lisa. "They are big as well; at least, with those who need their product in their supply chain. The difference is the target audience. Both are very successful, but they have different buyers. In both cases, the branding strategy is to make sure the right people know about them."

"I guess I am not their target buying persona." Jacob laughed.

Lisa nodded. "It's all about building the right strategy for those who can benefit from the value they are offering. The brand story needs to precede your arrival in front of a prospect. Then, every new and satisfied customer will carry the value proposition to others in need of their solution," she said. "When your mission is to address real human needs at their source, then brand should not be second to anything else. It is the ultimate expression of value, validated by those who benefit from your innovations."

Lisa spent the rest of the afternoon teaching Jacob how to understand symbols of innovation and exploring the depths of a mature marketing strategy and its role in the success of a business.

"Now, check this out, guys. Look very closely." Jacob snapped to attention, but Rhonda waved over to Tim to make sure he wasn't missing her presentation. She had invited Jacob and Tim over to her place that night to show them something new and exciting that she was working on: the perfect cherry on top of Jacob's incredibly day of learning.

"Whoa, this doesn't look like something your company is working on," said Jacob. "Another side project?"

Rhonda beamed. "Yes, in my spare time, I've been working on something special, and I wanted to get some feedback."

"I'm not great at jewelry, is that a ring, bracelet, or a necklace?" Tim asked, laughing. Jacob couldn't tell, either. She handed it to him to study closer.

"It's your choice. Until now, we've only ever had bulky virtual assistants," she said. "But I've been able to engineer something more stylish that people can put on themselves to express themselves in their own way." She gestured toward the device, now in Jacob's hands. "So, what you're holding there is the bracelet that than can extend to be a necklace or retract to become a ring."

"Yeah, this is stylish," said Tim. "I would have no idea it had any tech in it." He paused, raising an eyebrow. "It does have tech in it, right? You're not suddenly pivoting into jewelry design?"

Rhonda rolled her eyes. "Yes, it has tech, thank you. And very good tech, I would say. It has near-field communications, Bluetooth, and wi-fi; everything you need for connectivity to another smart device. It's seamless for the end user, and it has voice recognition input to provide that hands-free experience people have been looking for."

"Stylish and innovative," said Jacob, impressed. "Fascinating!" He turned the device over in his hands, finally finding the well-disguised port. "That's for the inputs and connectivity. I know you've been wanting to do something with the outputs. Any floating keyboards?" he asked.

"You're reading my mind," Rhonda said, grabbing the bracelet back from Jacob. "Check this out." She slipped it on her wrist and pushed a button Jacob hadn't even noticed before. Suddenly, a virtual touchscreen with multiple button selections hovered in the air in front of her. It looked like a smartphone home screen floating in the air that responded to her hand input.

"Oh, wow," Tim shouted. "You've got to be kidding, Rhonda! You didn't do what I think you just did, did you?"

Rhonda just leaned back and smiled. "Gentlemen, I've done it. It works. We are one step closer to the era of spatial computing. We will free users from the shackles of the physical keyboard and external monitor."

As Jacob watched in awe, something clicked. *This* was the type of work he wanted to be doing. He wanted to explore innovative, single-solution products that would really make a difference in customers' lives. Ideally, he wanted to do all that with the two people in this room: his best friends, and the smartest tech team he knew. At a huge company like Omnify, he and his solutions would never be anything more than cogs in the wheel.

That night, he sent an email to Dan asking for a meeting as soon as possible.

CHAPTER 14

OVER AND OUT

FRIDAY EVENING, JACOB AND ROBERT WALKED THE RIVER-walk, and Jacob filled him in on his latest thinking and his meeting with Dan.

"Omnify just isn't where I need to be right now," he concluded. "And I told Dan as much. He's working on an exit plan for me as we speak."

"I'm proud of you for making that move, Jacob," said Robert. "That takes guts."

"Thanks," Jacob replied. "I'm nervous, though."

"What about?" Robert asked.

"I loved developing TaskCore and all of the user-experience strategy and planning that came with it, so I'm not worried about coming up with a new product. That will be the fun part. But I just keep having flashbacks to how tough it was to get in front of enough prospects. We never could make TaskCore stand out when we just had startup resources behind us."

Robert thought for a moment, and then whispered, "Hey Jacob, come closer." Jacob laughed and made a show of leaning in to listen. "I see now that you got it all wrong. You didn't have a marketing problem, son; you had a business model problem. You had the right solution to the right problem with a business model incapable of joining the two."

"What do you mean?" asked Jacob.

"Say I'm your prospect," said Robert. "Maybe my company already has everything we need. Maybe we've come across you already and didn't see your solution addressing a problem we had because the message was confusing. Maybe it's a bad time and your brand lacks the persistence needed to be there when the time is right. Maybe it's the wrong product for their use cases. Maybe it's not a good pricing fit." Robert paused, and Jacob let this all sink in.

"There are a million reasons a product might not be the right fit for the buyer," Jacob said.

"Exactly," replied Robert. "But when we're getting ready to launch a new product, we have got to figure out what the right market strategy is so your solution will find the right problems at the right time. We have to find out how big a market we have and how many people are willing to spend money on this solution. And there's a lot more out there than you haven't seen before; with Lucera or Omnify. Maybe next time around, you'll have potential to grow your business rapidly. Or perhaps your business will grow in proportion to the needs of that potential future market share."

Jacob's mind was spinning with questions. Was that where TaskCore had gone wrong? How many people out there had the problem that his solution could fit. And conversely, did he build the right product to solve the problem? Or was TaskCore the wrong solution? Or did it used to be the right product before Omnify changed it? Or did most people just didn't know the right solution existed."

As he was thinking about it, his cell phone rang again. It was Dan. He needed an answer.

"Jacob, are you still serious about cashing out?

"Danielle, Jacob, thanks for inviting me out for lunch. I really appreciate it," Angela said.

"It's our pleasure," Danielle responded.

"I thought it'd be a good idea for you to meet Danielle as you evolve your career. I know you have not had success winning me as a client, but I've learned a lot in the last year that makes me empathetic to how tough it is to be in the field selling." said Jacob. "Danielle is the best salesperson I know."

"Well, I don't know if I'm really all that," said Danielle with unnecessary humility," but I'm happy to help out however I can. How are things going for you right now?"

Angela said, "Oh girl, it's been tough, as Jacob could attest. I feel like I'm always pestering my prospects to get their attention. My company gives us some training and a quota, then sends us out there to hit these numbers however we can. It's tough sledding."

"I remember those days. I got about ten years on you, though I know I don't look it." Danielle winked, and both women laughed. "Yeah, I've been in the field for quite a while. It can be tough when you're just getting started," Danielle continued. "But you shouldn't be discouraged. One thing I had to find out the hard way is that getting a prospects' attention and then developing a relationship that turns them into a client is a long process. And I learned early on that sometimes it's about getting lucky enough to happen to be around at the right time."

"Don't I know it." Jacob shook his head, thinking about meeting Nia and Michael in line at the food truck.

Danielle continued. "I had to find my way in this industry to find a company that provided value I really embraced. It made a huge difference. I love what we sell, but I had a lot of positions where I was trying to sell things I didn't believe in, and it really discouraged me."

Angela looked surprised. "Really? Tell me more."

Danielle went through some of the rough patches in her career and all the different places she worked. Jacob was glad Angela was getting this nuanced perspective of how hard Danielle had worked to get to where she was today.

"When I found the right place with the right solutions, I just grew there. My management team supported me in looking for long-term growth with my customers, which allowed me to become a true partner to them instead of just a salesperson. It made all the difference in the world. I no longer felt like I was just looking down at a list trying to sell everything I possibly could."

"That's what I feel like I'm doing most days," Angela said ruefully.

"That's definitely what you were doing the first twenty-five times you called my cell," Jacob teased gently, relieved when Angela laughed instead of looking offended.

"We all have to start somewhere, I guess," she said with a shrug.

"It helped that it was services that I was drawn to more than products," Danielle added. "Services allowed for more collaboration."

"That's an interesting point," said Angela. "I've only ever sold products."

"Well, based on your career track, Angela; not to mention Jacob's good word; I think there are some opportunities out there for you. Let me look at your resume and then let's see what we can come up with for a plan," Danielle said.

Angela seemed relieved. "I really appreciate that, Danielle. I'm glad I had an opportunity to open up and really talk with someone who understands where I am coming from."

Danielle and Jacob lingered a few minutes after Angela had left.

"Do you really think she could have what it takes?" asked Jacob.

"I do." Danielle nodded. "She's clearly sharp, and it's obvious she has the drive to get better. You can tell she wants to be passionate about her work. With the right offering and the right team, I think she could be great."

"That's good to hear," said Jacob, relieved. "Because I think I may have an opening for a salesperson soon."

In the small conference room of Dan's office, Jacob's lawyer reviewed the document one final time, then slid it over to Jacob, nodding that it was okay to sign.

Jacob looked around the room, acknowledging Dan and Tom, then took the pen from his lawyer's hand and signed and dated the agreement.

And with that, he'd cashed out of Omnify. He passed the form back to his lawyer, stood up, and shook Tom's hand. "Tom, take care of TaskCore. It's in your hands now." Then he turned to Dan and said, "Dan, I appreciate this opportunity. It's been a bumpy run, but I learned so much here about the challenges of running a large business. I'm looking forward to the next chapter. And I am so glad you spearheaded the efforts to stabilize TaskCore after all the integration issues with Synetra."

"I owed it to our clients to be better," Dan said. "You're the one who reminded me of that. We loved TaskCore when we bought it from you, but we just had too many competing priorities to do right by it. It was unfortunate that we didn't do better before we lost WellThrive as a customer. But it is now stable and delivering for other businesses in our healthcare division. Your original vision is no longer there; however, I know Tom is happy that he was able to deliver on the promise of reducing billing turnaround times for our clients that justified us acquiring TaskCore in the first place."

"I appreciate that," Jacob replied sincerely. "And I have no doubt that those like Nia and WellThrive will find another way to get their needs met."

"Well Jacob," said Dan, "I wish you nothing but the best of luck, and I look forward to seeing what you do in the future." Tom echoed Dan's sentiments, and with that, Jacob and his lawyer headed toward the elevators.

"Jacob, you're a free man now," said the attorney. "It's on to the next thing. Are you ready?"

Jacob smirked. "I've been ready. I've gone through what I needed to go through and learned what I needed to learn, and I'm very excited about the next chapter. I finally know how to turn a product into a functioning business. I have a clear vision of how to get the right solutions to the people with the right problems, and I'm now confident I can deliver on the brand promise."

CHAPTER 15

NEXTLINK

Jacob and his mom Sonia snuck in just before Nia's speech was set to begin, and they took a spot in the back row. Michael and some of the WellThrive's advisors were seated in front. She was about to speak at a prominent health expo, with doctors, nurses, surgeons, insurance professionals, spiritual healers, wellness coaches among the attendees. The theme this year was "The Future of Care: Human and Holistic." As far as Jacob was concerned, there was no one better than Nia to speak on that theme in an effort to help people to reclaim their best health.

There was a buzz in the air. Jacob knew from conversations with Nia that the world of healthcare was approaching a turning point. From alternative approaches to better nutrition to new technologies to more care delivery methods, Nia and her industry were on the cusp of changing the lives of millions and transforming our society.

The moderator introduced the founder of WellThrive Health to loud applause, and Nia opened her talk with a story.

> "When a little girl looks into her mother's face, she sees her future self. And that little girl and her future self, her mother,

> deserve as many quality years together as possible. I was a little girl who missed out on those years. I prayed for more time with my mother, but her life was cut short by declining health and compounding ailments. It should never be that way. There is so much we can do at the front end of life, way before we have to be over-medicated or rushed to the hospital. I know in my heart that my mom could have lived a longer life. My mom was my inspiration for helping others enjoy more time with their loved ones. Better wellness early and ongoing is better health for the future, and that leads to a life of thriving.
>
> Healthcare is too expensive for far too many people. Many patients' ailments go undetected and untreated. Getting coordinated care is too cumbersome and complex. Serious medical conditions are allowed to develop over a long period of time without being properly addressed. We can do better!"

As Nia paused for applause, Jacob saw her eyes flick to the back and a smile appear on her lips as she noticed him and his mom.

Nia closed her talk with a passionate call to action.

> "There has never been a better time to make transformative change in others' lives. We have more collective knowledge, learned lessons, novel techniques, breakthrough therapeutics, and advanced technologies than ever before. And we have to be willing to tackle the affordability challenge from the front end. We can all do better. The work that my company is committed to is making a difference, and so is yours. And we have so much more planned in the future. Join us to meet the needs of millions. Let us treat the entire person up front instead of the back end in the hospitals. Because that little girl deserves many more years with her mom. And her mom deserved more years of health and vitality. Let's be the change will all know is possible."

With that the audience leapt to their feet in an inspired standing ovation.

Jacob and his mom waited as Nia made her way through the crowd, shaking hands and chatting with other healthcare professionals, before she made it to the back of the room.

"That was an amazing call to action, Nia," Sonia said as she gave her a hug.

"I had no idea I would see you two here," Nia replied as she proceeded to hug Jacob.

"I had to see this in person," Jacob said, beaming at his friend that was taking on such a big challenge.

Jacob spent the evening after the expo at his parents' house, brainstorming. As he worked the sticky notes on his whiteboard, his father came over and put his hand on his son's shoulder.

"It looks like you're stuck there".

"Yep," Jacob replied, letting out a huff of air. "WellThrive left Omnify when TaskCore went downhill, and I don't blame them. But I hate that that leaves them without a way to address their insights-sharing challenge. I'm using them as the avatar for the new solution I want to develop."

Jacob's dad said, "Son, there's more than one way to get this done. Obviously, TaskCore was a great product and a great solution. But it's not the only solution. As you know, Jacob, often as I'm designing new buildings, I'm working with a lot of constraints, both internal and external. Sometimes we're working with limited budgets so we have to think strategically about building materials; in other cases, we have physical limitations that hinder our ability to do what we want to. There may be awkward plumbing and electrical systems we must work around, or people may be using one part of the building while we're building the other. There are always constraints."

"I get it, Dad," said Jacob, "but I'm not following how this applies to Nia and WellThrive."

His dad paused, then continued. "Okay, here's an example that more closely matches your challenge. An upcoming retailer's flagship store was designed entirely around a frictionless, cashierless checkout experience that used an automated system with ceiling-mounted sensors, RFID shelf tags, and AI software that tracks and charges customers as they leave."

"I'm with you," said Jacob.

"The entire store design, from layout and lighting to IT infrastructure, was built around this single-vendor technology. But ten months after opening, the vendor discontinued the platform, ending technical support, updates, and customer access. With no traditional checkout counters or fallback system in place, the store could no longer process transactions as designed. Short-term workarounds like mobile checkout would disrupt the customer experience and increase operational strain, while long term solutions would require costly redesigns of the physical space, new system integrations, and retraining."

"So, what did you do?" Jacob asked.

"We talked with the owners and developers, and we all agreed that the design of the physical store must be in service of the customers. Their experience was the central focus. The job was to facilitate a different experience when the customer walked into the store. The owners went out and found a vendor with a mobile self-checkout app that they could launch quickly instead of redesigning the physical layout of the store. Customers could scan items with their phones using barcodes or QR codes, pay directly through the app using digital wallets or saved payment methods, and verify purchase with a digital receipt shown to staff at exit," his dad replied.

"It worked because it leveraged existing infrastructure, preserved the "tech-forward" feel of the store, allowed for quicker deployment than full system replacements or physical redesigns and allowed for partial automation to maintain the thin staffing. This approach helped bridge the gap between full automation and manual fallback while buying time to implement a more permanent solution: replacing the core system with an alternative vendor whose solution was market ready within a years' time frame."

"So what you're saying is WellThrive needs a short-term fix while they look for a new solution that leverages its current infrastructure?"

"Exactly," said Jacob's dad. "Tell me a little bit more about this. How was WellThrive using TaskCore before you sold to Omnify?"

"They were using it as their knowledge management system to supplement their other tools," said Jacob. "At the time, they didn't have an electronic health record system, and they were trying to capture all the information their specialists and doctors were gathering from patients. If they had one recorded history of certain symptoms and had created a wellness plan that worked, they wanted to make sure they captured it so the next specialist could leverage that information to more effectively treat similar situations. Most common health records just don't work that way. So, we built our knowledge management system and a schema around all the different workflows from the different specialties, from standard physical family medicine to yoga to wellness to nutrition to the spiritual breakout sessions they offer their members. As they collected and captured all this information, we were using the fields to actually do correlations on the back end to flag patterns."

"Okay," Jacob's father said. "So, with TaskCore out of the picture, how does their knowledge-sharing work now?"

"Well, now they'll go back to the standard tool set they had before," Jacob said.

"That was a while ago, son," his dad prodded. "Has anything changed in their environment?"

As Jacob thought about it, the pieces started to click into place. "Actually, you're right, Dad. Things have changed. They did go out and get an EHR system. That's the stop-gap solution. At the time, they were just going to use certain functionalities to cobble together a solution, but I wonder if we could leverage that in a new and different way."

"You probably could," Jacob's dad replied. "If we can't use one material or one design aspect in a given situation, we'll make adjustments. Now, obviously, if we're adjusting where a support beam goes or where the windows go, that affects a lot of other things. But I'm sure, Jacob, with their new EHR system and any other process changes it's sparked, you could come up with something that solves their problem in a new and more lasting way."

Jacob looked at his dad and said, "You know what, I think I have an idea. Excuse me, dad, I think I need to get this down on my board as soon as possible."

In what felt like no time at all, the sticky notes on Jacob's whiteboard had turned into a full-fledged product vision. Robert had helped him develop a business plan that would support the right-sized company, the right number of people, and the right number of products to meet his goals and live comfortably while fulfilling his mission. Robert had cautioned him about chasing his tail in search of solving every problem he saw. "That mindset leads to a path of perpetual burnout, trying to be everything to everyone," he'd said, encouraging him to scale at a pace that was comfortable for him and his team.

When Jacob presented his vision and his business model to Rhonda and Tim, they were on board. They moved quickly: the three friends founded a new startup, named it NextLink, and started building their new flagship product, called SmartFlow.

Jacob admired their brand-new office, a larger space back in the Innovations Hill complex on a different floor than before, this one facing the inner park area. He was proud of what he and his friends had spun up with everything they'd learned after the sale of Lucera.

He watched Rhonda and Tim working up a SmartFlow model, discussing some of the features coming out in the next iteration.

"What are you two doing over there?" Jacob chirped. "Causing trouble? You got it just right, so don't mess it up now! I think our customers are going to like it just the way it is."

"Come on Jacob, you know we can always make this thing better. We just need to fine tune a few elements," Rhonda joked.

Tim replied more seriously. "Hey, our approach is rock solid, but we have to keep on top of our game. Our clients' challenges are always evolving. Our solution has to continually adapt as well to support the way our clients work," Jacob had been nervous about product iterations after what had happened to TaskCore, and he appreciated Tim keeping him on track. "I think this update will be a big help to the user experience. Take a look at this."

Jacob circled around and zeroed in on this new element that he hadn't seen before.

"What do you have going on right there? Your doing, I assume, Rhonda?"

"Ah." Rhonda smiled. "Yes, this new element right here opens up a new capability, leveraging the new hardware embedded into the wearable glasses to allow the customer to complete the entire new user intake process without having to stop and go back to their laptop. There's a secure transmission between the smart glasses, user wearables, room sensor nodes, and in-air screen projectors, which allows us to transmit data over any network. So even if you get disconnected from the network, it'll reconnect when it gets into close proximity with either near-field or Bluetooth communications. It works seamlessly."

"And before you ask," Tim chimed in, "all the compliance and data privacy features are already accounted for. No matter how users set their privacy needs, we will maintain feature parity. We've got this thing locked up tight. All the different elements are in place, Jacob. This is exactly what we've been researching and planning and looking for since the Lucera days."

Jacob smiled. When he and his friends had started to think about all the ways they could change the world with Lucera, he'd taken for granted that they would always work together. When they went their separate ways, he didn't know if they'd ever get the band back together. He was so glad to have them back as partners in this new business venture. Jacob felt prouder to be working with his two best friends than he had any other time in his career.

"I'm lucky to have you two," he said. "Come on guys and give me a hug. This is fantastic. This is a dream come true for me."

It wasn't just Rhonda and Tim who'd joined Jacob at NextLink. Every startup needs a sales lead, and following through on the prediction he'd made to Danielle, Jacob gave Angela the position. She was still green, but on this new, smaller team with a single focal point that everyone was passionate about, she was thriving.

"So, Jacob, is this how you want this slide deck to look? I think I get what you need, but is this the image that the client is going to be looking for when we get to this particular functionality?" Angela inquired. The image wasn't focused on tech specs but on outcomes. Angela was so accustomed to tech-first selling that Jacob knew this still felt strange to her.

Jacob turned around and looked across the table. "That's it, Angela. You hit it spot on. We want to make sure the client knows the potential upfront, that our solution overcomes each of their pain points or challenges. When you reach out to someone on an expo floor, a sales preview, or even on the street, you can quickly pivot to this one slide right here. It's our value proposition in a nutshell."

Angela said, "Wow, this is amazing, Jacob. I never would have had to hound you on the phone and online so much if I'd approached sales this way years ago!" Jacob laughed, and Angela continued. "This feels so much better. I am not afraid I'm a burden on my prospects, anymore, because I really care about helping them, and I know just how to do it. I feel proud to say that I'm here to help as opposed to trying to steal a few minutes of time. I really appreciate you bringing me on and helping me out."

"Don't worry, Angela," he replied. "I've felt your pain over the years. I didn't understand exactly how to verbalize it, but I understood how hard it was to make those calls and run as many scenarios as possible through your mind at one time. But with everything in alignment around serving our clients, I think you're going to really thrive in this environment."

"You think so?" she asked. "I'm still a little nervous about this first sales call."

Jacob nodded. "I've seen your practice sessions. I've seen how you present. And I know right now you're backed one hundred percent by what we're positioning. You're not selling to prospects; you're making them aware of something that can help them address their problems and meet their needs. They're going to thank you for it," he said. "And besides, with you and the sales team being strategic about who you're going to go after instead of cold-calling anyone and everyone, it should be easier to find better prospects who are primed to really hear this message."

Angela looked relieved. "You're right. I've got this," she said with a smile.

"I'm happy to have you on board, and I know you're going to do very well here," Jacob said.

Jacob knew that, this time around, he wanted to be sure his message was hitting the right notes in the right places from the get-go; and especially as they started to deepen their focus within the healthcare industry. Fortunately, Lisa, the branding expert Robert had introduced him to, was happy to consult on that process. She and Jacob, along with Rhonda, Tim, and Angela, gathered in the conference room for a strategy meeting.

"Let's talk some more about the roadmap and the future use cases for SmartFlow," Jacob said, kicking off the meeting. "This initial clinical focus is really exciting, but as Angela has mentioned before, there are many possibilities for our solution to help improve peoples' experience addressing their health needs."

Rhonda nodded. "Yes, virtual in-air screening that our glasses and bracelet enable between any two people is just the tip of the iceberg for what we can do in healthcare."

"Yes," Tim agreed. "The way we've developed the initial stack allows us to be pretty modular. We can use these in different areas of the health field, and I would think there's a market for it. With the aging population and more complicated health ailments, there is a huge push on for improving patients' lives."

"Yes, I like this," said Angela, and Jacob was pleased to see her feeling confident enough to speak up. "I do want to make sure we stay focused in our messaging, though, even within different areas of wellness and health, so we don't dilute our value as we bring SmartFlow to the masses."

"You're exactly right, Angela," Lisa chimed in. "The value proposition has to be consistent from product development to marketing to sales in the field. So, the way we're branding SmartFlow is as a new way to connect people in the health space, freeing up everyone's hands and allowing more effective collaboration among providers and with patients. The connection aspect is universal enough that it would hold up if NextLink were ever to expand to other industries, but the framework is unique enough to healthcare to let us stand out in various traditional and new age healthcare environments. Angela, you will have a leg up on competitors when you present SmartFlow to our prospects."

Jacob was thrilled to hear all this. "I like that too. I think we know we need to avoid some of the pitfalls I saw at Omnify when we tried to be too many things to too many different industries. I'm glad we're on the same page, and I'm happy with this emphasis with proving our value first and foremost. And we'll do all of that with laser focus on improving patient outcomes in healthcare." He turned to Lisa. "What do you think?"

"I think you're spot on," Lisa replied, "Our core messaging around this solution will resonate very well with audiences, as will our storytelling about all the different healthcare applications. We won't be confined to conference booths or online ads, though of course we'll do those. But I'm confident this is a product that will spread by word of mouth, too. People will try it out, see the impact we're having, and tell their friends and colleagues in the industry how effectively SmartFlow transformed their workflows and patient outcomes. This type of solution is something that the market has been looking for. We just have to let them know we exist."

Lisa liked everything she heard and saw from the team. However, she knew the next step would be critical to all of their plans. She turned and looked at Jacob, "You know what you have to do now."

Rhonda chimed in as well, enthusiastically cheering him on, "Go get us some more money Jacob!"

Jacob's next big task might be his most important. There was enough funding available to get NextLink off the ground. Now it was now time for the type of outside funding necessary to enable his new business model to be successful. He positioned the company to be attractive to the right type of investors. This was a place Jacob shied away from before. But he was in a different headspace now with more confidence. The next phase in the development of NextLink was at stake.

CHAPTER 16

MOVIN' ON UP

ROBERT WALKED AROUND THE TABLE IN THE RECEPTION area and gestured for Jacob to come on through. He had set up a meeting for Jacob with Thomas, a friend and business partner who led a venture capital investment fund investing in young start-ups solving real customer problems, with promise for good growth. Robert had told Thomas all about NextLink, and he wanted to hear more.

After introductions and greetings, Thomas said, "Well, Jacob, tell me what's happening at NextLink."

"As Robert may have told you," Jacob said, "SmartFlow is the next evolution of my work designing solutions for human problems. I've had great experiences learning what it means to bring a good solution to market effectively. Specifically, I've come to understand: sometimes the hard way, that it's not enough just to have a great product or service; you have got to have a great business model, too." He looked over at Robert and smiled. "With mentorship from Robert and the experiences I had both in my first startup and at Omnify Technologies, I've gone through the growing pains learning how to design the right solution and build a viable business model around it."

"And where has all that learning taken you?" asked Thomas.

"I've reunited with my Lucera partners, also my childhood friends, who have great experience in engineering and coding. I've got a new team that's laser-focused on value-based selling, and I've got a clear idea how to market the solution. I think we have all pieces in place, Thomas. The last missing link is the funding to power our go-to-market strategy and sustain our product research. In the past, I was hesitant about taking outside money, which is one of the reasons why I took the journey to Omnify rather than bring on VC for Lucera. But I learned that selling to a larger company is just as likely to mean giving up control; and probably more of it. So, this time around, I'm looking further into venture capital to find the right fit."

Thomas said, "This all sounds good, Jacob. Robert gave me a lot of background and I'm really impressed with what I've heard so far. My venture capitalist fund is a little bit different than others. Our team really does look forward to solving some of mankind's most challenging problems. Unfortunately, in the past we've invested in a lot of technology companies who have promised to actually change how things work but not followed through. Even though we may have gotten a good return on our investment, we didn't ultimately have the impact we wanted to. I'm really excited about what I've seen here with SmartFlow. Why don't you take me through the solution?"

Jacob handed smart glasses over to both Robert and Thomas. "Let's experience the technology firsthand while we go through some use cases."

For the next hour, Jacob showcased the capabilities of the smart glasses, bracelets, and display projectors. They experienced the inline display, the voice-activated assistant, AI-assisted data capture, and the workflow engine.

Jacob could tell Thomas was just blown away by what he was

seeing. But he hoped the investor was even more blown away by the use cases and the human aspect of SmartFlow.

"I'm sold," said Thomas at the end of the demonstration. "Robert, I really appreciate you introducing me to Jacob. Jacob, you'll be hearing from me and my team here very shortly. I'm looking forward to working with you and NextLink."

The next day, still riding high from his meeting with Thomas, Jacob joined Angela on a Zoom call with a new prospect. The conversation turned to the topics of AI and big data in the healthcare industry.

"We are very concerned about how the use of artificial intelligence might compromise the recommendations we make as health professionals. We are continually asking technology to make decisions that have real patient impact. Our lawyers are concerned about where the liability falls if we provide the wrong care because some large language model didn't understand a patient's context. How is NextLink addressing the dangers of artificial intelligence running wild?" the prospect asked.

Angela replied, "NextLink takes these concerns seriously, as we are staunch advocates for the ethical and responsible use of artificial intelligence. We've thought about this a lot. We don't see SmartFlow as a gadget that uses shiny new technology or artificial intelligence in particular just for the sake of using it. And we certainly don't see it as some sort of replacement for humans. In fact, humans are at the heart of what we do, what you do, and what people in the market do. We are focused on giving the humans *we* serve a tool to make their jobs easier and better empower them to help the humans *they* serve."

"I've been accused of being a UX fanatic more than once," said Jacob with a laugh. "And that's okay with me, because I design for the user experience, with humans at the center. And I know there are certain jobs that providers don't have the bandwidth to do and organizations don't have the capacity to hire out, so we're left with process gaps that negatively impact the patient experience. So, what we do at NextLink is identify those gaps and use technology to fill them, amplifying frontline workers' capabilities; never undermining. Our technology is designed to aid human judgment, not the other way around."

The prospect nodded. "I like that philosophy."

"As you look to consider using NextLink's technology," Angela said, taking the reins back from Jacob, "know that you and your people inside your organization will be looking to enhance capabilities across the board. And as you've seen in the documents we've looked at today, which I'll send over to you, we've put more thought into this than many others introducing emerging technology into the healthcare space."

"That's all good stuff," the prospect said. "So, shifting gears, another hot topic is data rights. Who owns it? Us or the user? Or the third-party platform that helps us manage it? What are your thoughts on that, Jacob?"

"I'm glad you asked about that," Jacob responded. "There are so many companies out there who don't respect the privacy of people's individual data. In fact, they treat their customers as the product. I've been preaching that I know the market is going to shake out for those who care about privacy and value their customers' rights. We must be responsible for educating our own customers on how their data will be used and what their rights are. I think this is going to be a differentiation in the marketplace as more and more consumers demand more respect for their own data."

After several more questions that Jacob and Angela volleyed

between them, the prospect said, "We are excited to be working with NextLink. We feel more comfortable with how your tech will impact our employees and our corporate culture than we do with the other guys on the market."

With that, the meeting came to a close, and NextLink had yet another excited prospect. Angela was proving her worth and more, and Jacob couldn't be happier.

Weeks later, Jacob and Angela were having lunch with Nia and Michael.

"You know our challenges, Jacob," Nia said. They were celebrating a successful onboarding after WellThrive had decided to integrate SmartFlow with their EHR system.

"We felt like our technology system's interface was a hinderance," said Nia. "Our members felt rushed and not really listened to. They usually gave us the benefit of the doubt, but it was causing unnecessary friction. We wanted our advisors to be freed from the tether of their laptops and devote their human attention to the people in front of them."

"SmartFlow has put us on the right path, and we are looking forward to its future capabilities. We also were hoping out the box it would address the insights sharing aspects that TaskCore had. Our current workarounds will need a better solution in the near future." Michael said.

"I know," said Jacob. "I hate that you had that experience and we are committed to restoring those TaskCore-like insights sharing capabilities for you soon in a new and novel way. And for some of your new challenges, I had WellThrive in mind specifically as I put together the vision for SmartFlow to improve your advisor-member experience."

"We could tell," said Nia. "I just love the way you and your team listen to what we we're experiencing and think carefully about the solution. We immediately connected with you on SmartFlow. It's a real partnership, not just a sales pitch. We just feel very comfortable working with you guys, and we really appreciate what you brought here."

Jacob just smiled and said, "This is why we built the new business model. Perfectly sized and partnered with the right type of VC funding that is aligned to our mission. We're here to make sure the solution works, and not only after we sell it, but for years to come." Jacob was gratified that this focus had paid off. "I'm always glad to hear this feedback from customers and especially the new ones that we just got through onboarding. I'm loving what I'm seeing from you guys, and you should give yourselves a hand as well."

"Jacob, you're being too modest," said Michael. "Let me just say to everybody around here, I'm glad you went through the pain. You listened and thought deeply about our problems. You took that innovator's mindset and didn't just build a great product; you figured out how to build a business out of it."

"And we're proud to be one of your customers," Nia added. "Like you said, we had a problem looking for a solution, and then you built that solution. And as we grow our business, we will continually have new needs come up that require a solution to address it. I am confident you will be able to keep up with us. Thank you very much, Jacob."

"You are welcome. And I am grateful that you gave me another chance after all these challenges," said Jacob. "And grateful I had Angela here to help. She's been an excellent partner to our clients."

Nia raised her water glass, smiling. "Angela, you are with a great team there. And you're a great ambassador."

"Thank you, Nia. I've found a home here at NextLink. I am learning so much from Jacob and the others."

With that, the talk turned to WellThrive, and Nia and Michael excitedly shared their expansion plans. Their members were increasingly referring their friends and families to WellThrive as the transformative nature of quality holistic healthcare was showing tangible results. Patients who had been experiencing chronic mental and physical issues for years were now reclaiming their power to improve their whole selves; physically, mentally, and spiritually.

Healthcare's future was bright thanks to the hard work that Nia and WellThrive had done over the last six months. Jacob was honored to have contributed to that success through his ability to leverage technology to improve the human experience.

"Your marketing team really wants the prospect to know how reliable their system is...how much due diligence went into the governance structure...the number of other clients they have; which could all be interesting proof points," Jacob explained to the panel on CNBC's *Lunch Power Hour* show. "But the real question is whether you can show those prospects you're addressing a real problem and not just creating market noise. What are you solving for? You really got to nail that down."

Jacob was relieved he'd finally made an appearance on this show. For so long, he'd thought achieving this goal would be proof he'd succeeded. He was beyond proud of himself and his team for getting here, but he was even prouder to be able to share what he'd learned; the way Robert had always done for him.

"There are so many problems in the world," he went on. "Your prospects are in search of help. Sometimes they don't know they need it. Sometimes they don't know how to ask for it. Some don't know what that help looks like. Others can't conceive of what a useful solution might be. That is our job as innovators and solution providers: to connect the dots from what we've built to how it solves those pain points."

"So, Jacob, it sounds like you are speaking from personal experience" the moderator of the panel chimed in.

"Yes, I am," Jacob responded. This is one of the reasons I'm excited to speak on your show today. It is my responsibility to spread the word to those out there who don't believe the help exists. I want to tell them I have listened to and learned from their problems, and I've built a solution that tackles them head-on."

Thirty minutes later, Jacob wrapped the interview and headed into the green room where his mom and dad were waiting for him.

"I told you your time would come, Jacob," his proud Mom teared up as she hugged him.

"Jacob, I am so proud of you son," his father beamed.

Jacob hugged both his parents. "This is the moment I've been dreaming of."

EPILOGUE

THREE MONTHS LATER, JACOB, RHONDA, TIM, AND ANGELA gathered around the table in the conference room for a brainstorming session, or jam, as Rhonda liked to call it. They were futurecasting what was ahead for NextLink and their mission to revolutionize patient care. They were very pleased with the value they were bringing to the market and their new clients. The value was obvious, and the clients were happy in both the surveys and reported direct user impact.

And the icing on the cake was a large expansion of the contract with WellThrive.

"I just came from WellThrive," Jacob said. "Nia and Michael were discussing some of the feedback their members gave them about transitioning from an in-clinic visit to home. You know me, I was taking mental notes." Jacob laughed. "Patients said that after they left the wellness session and went home, there were several manual handoffs with the home web portal, and they were having to manually enter some of the journals and vitals. The members were jotting down their own notes to enter into their online charts. It got me thinking about more ways we could help WellThrive provide more value for members after they leave the clinics."

"What do you got?" asked Tim.

Jacob proceeded to login into Wireloom, the wireframe and design tool he saw at the *Future of UX Conference* last year. It was his new go to place to transfer his sticky notes brainstorming ideas to a virtual design canvas.

"After I left, I sketched out the member journey," Jacob said. "I'll bring up the Wireloom screen here and everybody put their glasses and their bracelets on so we can walk through it together."

And all four of them put on their bracelets and entered the virtual world. "With WellThrive and so many others like them, what happens when patients, users, and members go home?" He highlighted all the data and tracking members were responsible for outside of the clinic. "And then look at all that happens when they get back in front of the professional. They spend all this time recounting the information so it can go in the provider's system."

"Rhonda, these smart glasses and virtual displays are just the beginning of the possibilities for leveraging augmented reality and virtual reality for information sharing, status reviews, and explanations of follow-ups. What if members could use them at home to log their data in real time to keep the clinic system up-to-date with their statuses and flag any changes? What if they could directly ask follow-up questions rather than going through multiple channels and portals to get the provider's attention? What if their in-home experience was just as seamless as their clinic experience?" "I like where your head is at," said Rhonda. "What would the display look like for that?"

Jacob switched to the next slide, showing a UX mockup of a home dashboard. "Very intuitive, avoiding all the manual in-person tasking of keyboarding and relying on employees' memories and standard processes to do day-to-day operations. This is that virtual command pad that's the overlay for the end client session. It's intuitive, graphical, and context-sensitive to both the doctor and the patient as they interact with each other."

Both Rhonda and Tim were on the edge of their seats.

"This would be the SmartFlow Extension for Home," Jacob concluded, grinning with excitement.

Rhonda said, "But Jacob, the home? That would take us into a new device area. We just don't have that type of hardware capability now. And what about the increased support requirements to handle home users? That is a tall ask."

"I know Rhonda, you're right," Jacob responded. "That is, if we tried to do it ourselves. Our business model supports the right type of partnerships. We are talking about more hardware and end user support. Our solution scales at a higher level. If we want to be able to help clients like WellThrive expand their impact to their patients, they need the ability to interact outside their physical clinics. From tracking health vitals, to journaling, to symptom look-ups, patients need more connection in between their clinic visits. I have been talking with Dan and Tom from Omnify about a strategic partnership."

Tim gasped. "What? I thought you were done with them."

"They have specific capabilities we can leverage," said Jacob. "Patient needs and their problems aren't confined to a clinic. Our business model can support this, and Omnify likes what I presented." He beamed. "The needs of the human aren't bound. Neither is our solution."

Angela asked, "And what about Rhonda's concern about the support side for home users? That's not in Omnify's wheelhouse."

Jacob smiled. "I've spoken to Danielle, and Integra is interested in partnering with us to assist."

"Well look at you," Rhonda said. "You're becoming more a businessman every day." Jacob had learned a lot about strategic thinking, sure, but he was grateful that Robert had helped him map out partnership strategies to support NextLink needs.

"As Robert taught me, it takes a lot more than a good idea to help people. You need a business to bring solutions to those who need them."

"I am so proud of you, man." Tim patted Jacob on the back.

"Watch out, guys. There are more problems that are about to be on the run from our solutions."

AFTERWORD

Innovators like Jacob design solutions to solve problems for people. They build products and services to be the vehicles of value. They build businesses to deliver those solutions to the market. Innovators must keep their focus on people's needs.

When we keep people as our guiding light and listen to their problems, issues, and challenges, we will build and deliver the right solutions to the people who need them. We make money when we deliver value to people.

Keep people at the center of your efforts. Innovation is valuable when solving problems, alleviating pain, and completing jobs for real people. That's the true meaning of value realization.

At this stage in Jacob's journey, he is wiser. He's built the right business model and assembled the right team, and now he's well positioned to help solve some big problems.

But Nia, who has set out to improve the physical and mental well-being of people, is left with her company's needs only partially met.

By the end of our story, she is making do with workarounds, however she's optimistic that Jacob's new solution that's been recently integrated with their wellness systems can match the functionality of his old solution and keep up with the evolving needs.

But she still has a lot of work to do to meet her goals.

If Nia is going to change the world, Jacob and team are going to have to deliver just the right kind of innovation.

It is time for both of them to realize some significant value!

www.ingramcontent.com/pod-product-compliance
Lightning Source LLC
LaVergne TN
LVHW020414220926
848484LV00018B/2576/J
* 9 7 9 8 2 1 8 8 6 0 1 8 9 *